CHANGE YOUR
BRAiN
CHANGE YOUR
life

BEFORE 25

Change Your
Developing Mind for
Real-World Success

Jesse J. Payne, Ed.D.

CHANGE YOUR BRAIN, CHANGE YOUR LIFE (BEFORE 25)

ISBN: 978-0-373-89292-1

The health advice presented in this book is intended only as an informative resource guide to help you make informed decisions; it is not meant to replace the advice of a physician or to serve as a guide to self-treatment. Always seek competent medical help for any health condition or if there is any question about the appropriateness of a procedure or health recommendation.

Brain Systems Quiz on pages 5, 168 and 263 copyright © 2014, Amen Clinics, Inc.

Library of Congress Cataloging-in-Publication Data

Payne, Jesse J. (Jesse Jeremy)

 Change your brain, change your life (before 25) : change your developing mind for real-world success / Jesse J. Payne, Ed.D.

 pages cm

 Includes bibliographical references and index.

 ISBN 978-0-373-89292-1 (alk. paper)

1. Mental illness—Physiological aspects. 2. Brain—Pathophysiology.

3. Neuropsychiatry—Popular works. 4. Mental efficiency—Popular works. I. Title.

 RC455.4.B5P39 2014

 616.8—dc23

                    2013038937

www.Harlequin.com

Printed in U.S.A.

Contents

Part III:
How to Change Your Brain and Change Your Life (Before 25)　163

Foreword

Your brain is involved in everything you do—how you think, how you feel, how you act and how well you get along with other people. Modern neuroscience teaches us that your brain is the organ of intelligence, character, personality and every decision you make.

When your brain works right, you work right; when your brain is troubled, you are much more likely to have trouble in your life. With a healthy brain, you are happier, physically healthier, wealthier and more successful, because you make better overall decisions. When your brain is not healthy, for whatever reason, you are sadder, sicker, poorer, less wise and less successful, because the organ that makes your decisions and runs your life is not working at its best.

Change Your Brain, Change Your Life (Before 25) is an extension of my original book, *Change Your Brain, Change Your Life,* first published fifteen years ago. I asked Dr. Jesse Payne to write it because of his ability to communicate with younger people, and as you will read, he has tremendous personal resilience and character. Jesse and I developed a high school curriculum around brain health that has changed many teenagers' lives.

The information and stories in this book are powerful and truly life changing. If you really learn to love your brain, the actions to take care of it do not become a hassle. Rather, they are just something you need to do out of love for yourself.

As you read this book, I can guarantee that you will never look at your brain or the brains of others the same way.

— Daniel Amen, M.D.

Introduction

I was twenty-one years old, sitting in the office of Dr. Daniel Amen, a medical doctor, a world-renowned and award-winning psychiatrist, a *New York Times* bestselling author and...the father of my girlfriend.

Even though I am more than a foot taller than he is, I felt intimidated as he reviewed the scans he had just taken of my brain. I had been dating his daughter for a little more than a year, and apparently, agreeing to having a detailed scan of the brain was a requirement for anyone who had dated one of his daughters for any significant amount of time. He said it was to make sure that everything "looked okay." Riiight.

It was like a twisted version of the scene in *Meet the Parents* when Robert De Niro sweats out Ben Stiller—except this was very real. Plus, this wasn't a lie detector test. This man could actually peer straight into the inner workings of my brain.

As he gazed down at the images and data from my scans, I saw his forehead wrinkle as he frowned ever so slightly. "Hmmm," he said. This was not a good start. "I see that you can be pretty stubborn and argumentative at times. Does this sound about right?" he asked.

I knew that my response would shape the rest of this conversation, and possibly the future of my relationship with the girl I had fallen in love with. And yet instinctively, before I realized I was doing it, my arms folded across my chest. "No. I wouldn't say that I'm stubborn at all." My reply had an undeniable tone of defensiveness.

A hint of a smirk appeared at the corners of his mouth. "Are you sure about that?" he asked.

Ugh. My brain had betrayed me.

It was the first time I saw a connection between my brain and how it was related to my thoughts, feelings, actions and behaviors. More important, it occurred to me in that moment that if my brain determined my behavioral tendencies, perhaps the power went both ways and I could have some control over how my brain worked. The prospect gave me hope for my future.

Before I met Dr. Amen or his daughter, my life had not exactly been a stroll in the park. My mother and paternal grandmother were illegal immigrants from Mexico, and my parents were both high school dropouts. As a child, I endured physical, emotional, psychological and even sexual abuse. My mother suffered a nervous breakdown when I was four, a result of her struggle with schizophrenia, delusions and hallucinations. My parents divorced when I was eight, and I spent much of my childhood moving from place to place, living on welfare and food stamps. I still remember the day my mother kidnapped me from my father and led police on a high-speed chase through the streets of Fresno, California. My father, for his part, struggled with ADD, depression and anxiety until he gave up and committed suicide.

Yet, while I saw many people in my life lose their way, I never did. Somewhere inside of me, a fire always burned. I knew that I wanted to better myself and my situation. I wanted more out of this one life I was given here on Earth.

I can remember when I realized as a young child that applying dedication, perseverance, critical thinking skills and hard work to every challenge I faced was my ticket to a better life. As I watched many of my friends and family succumb to drugs or wind up in jail, I put my head down and studied hard. I worked on building strong personal relationships and developing innovative, creative outlets for my talents. I knew all these things would allow me to live the life I wanted for myself, and for my future family.

It worked, too. I graduated from high school with honors, and I was the first person in my family to receive a college degree. I went on to receive a master's degree and ultimately earned a doctorate degree from USC. Today I am healthier, wealthier and happier than I thought I could ever be.

I've tried to dedicate my life to helping young people realize they do not have to be held hostage by their circumstances. As cliché as it sounds, I have tried to show others that if a poor Mexican boy with crazy parents can make it, anyone can.

I'm not saying any of this was easy. I too often found myself fighting the urge for rebellion and chaos. It was hard to resist the temptation to have fun in the present at the expense of the future. It could be a challenge staying focused on the right things and what was best for me. But by using many of the strategies I outline in this book, I was able to rise above the tougher side of life and achieve my dreams.

After that fateful day with Dr. Amen, who would later become my father-in-law, I began to work with him in the Amen Clinics, where I eventually became the director of education and human resources. I worked closely with Dr. Amen to create the Making a Good Brain Great high school course, which is now in more than four hundred schools across the country. The course teaches teenagers to better understand their brains and to do less harmful things to them. And I've spent the past six years teaching brain science to students in high schools and colleges around the country.

My hope is that this book will empower you to harness the amazing capacity and potential of your brain, which I have experienced firsthand both personally and in my research. The young, developing brain is a marvel to behold, and if you take steps to care for it now, you will lay down the foundation for success in your future. The most exciting part of this is that just about anyone can work hard to change their brain and change their life. I am a living testament to this.

I didn't use my past as an excuse. I became empowered to change my brain, change my life and become the outlier of my circumstances. You, too, can learn how to change your brain and change your life. Let's get started!

Part I

The Brain
Before 25

1 BRAIN FACTS

For a long time, the brain has been the redheaded stepchild of the human body. Think about what you learned about the brain in school. More often than not, any lessons about the human brain entailed a long list of vocabulary words and facts about brain structures, neurons and dendrites that were boring and difficult to grasp. I would also guess that there probably wasn't any explanation of what these various brain structures actually do and how they affect you in your own life.

If this sounds similar to what you experienced in school, then get angry and prepare to start a revolution. The fact that we are not taught about the amazing complexity of the human brain, the importance of helping it work right, and the connection between our brain and our life is insane. When you begin to grasp some of the practical and easily understood brain basics, you will be amazed to learn that you have the power to improve your brain and have a better life. By understanding the inner workings of your brain and how they relate to all aspects of your life, you'll begin to understand why you are the way you are, and why you act the way you act. Your tendencies, your struggles, your personality—all of these come from your brain. But it's not like you're just born with the brain you have and you've got that brain your whole life. Once you realize that, the amount of power and influence you have over the functioning of your own brain—and, by association, your life—becomes clear. This is when the magic happens.

Of course, before you become the master of your brain, it is important to start with the brain basics. The eight brain facts in this chapter may appear absurdly simple and commonsensical. Well, this is true, and that's the beauty of it. What you will find after reviewing each of these brain basics (information that's based largely on the work I did with Dr. Amen during my time at the Amen Clinics) is that you have probably never stopped to put it all together. Perhaps you've known this information all along, but when you couple it with the powerful program in this book that will help you change the way your brain works, things will begin to click. And you will undoubtedly give your brain the respect it deserves.

Before you read any further, take the Brain Systems Quiz, which was adapted from the work I have done with Dr. Amen, to give you better insight into which parts of your brain might be contributing to some of the frustrations and/or struggles in your life. Please rate yourself honestly on each of the items in the list. Follow the instructions at the end of the quiz to determine what your answers mean. After you've analyzed your answers, put the quiz aside. We'll come back to the results later on.

BRAIN SYSTEMS QUIZ

Please rate yourself on each of the behaviors listed below using the scale provided. If possible, have another person who knows you well (e.g., a parent, a significant other, a close friend) rate you, as well, in order to construct the most complete picture.

0	1	2	3	4
Never	Rarely	Occasionally	Frequently	Very Frequently

_____ 1. Failure to pay close attention to details; tendency to make careless mistakes

_____ 2. Trouble sustaining attention in routine situations (e.g., homework, chores)

_____ 3. Trouble listening

_____ 4. Failure to finish things; tendency to procrastinate

_____ 5. Poor time organization

_____ 6. Tendency to lose things

_____ 7. Tendency to be easily distracted

_____ 8. Poor planning skills and a lack of clear goals or forward thinking

_____ 9. Difficulty expressing empathy for others

_____ 10. Impulsiveness (saying or doing things without thinking first)

_____ 11. Excessive or senseless worrying

_____ 12. Upset when things do not go your way

_____ 13. Upset when things are out of place

_____ 14. Tendency to be oppositional or argumentative

_____ 15. Tendency to have repetitive negative thoughts

_____ 16. Tendency toward compulsive behaviors

_____ 17. Intense dislike for change

_____ 18. Tendency to hold on to grudges

_____ 19. Upset when things are not done a certain way

_____ 20. Tendency to say no without first thinking about a question

_____ 21. Frequent feelings of sadness or moodiness

_____ 22. Negativity

_____ 23. Decreased interest in things that are usually fun or pleasurable

_____ 24. Feelings of hopelessness about the future

_____ 25. Feelings of worthlessness, helplessness or powerlessness

_____ 26. Feelings of dissatisfaction or boredom

_____ 27. Crying spells

_____ 28. Sleep changes (too much or too little)

_____ 29. Appetite changes (too much or too little)

_____ 30. Chronic low self-esteem

_____ 31. Frequent feelings of nervousness or anxiety

_____ 32. Symptoms of heightened muscle tension

_____ 33. Tendency to predict the worst

_____ 34. Conflict avoidance

_____ 35. Excessive fear of being judged or scrutinized by others

_____ 36. Excessive motivation (e.g., can't stop working)

_____ 37. Tendency to freeze in anxiety-provoking situations

_____ 38. Shyness or timidity

_____ 39. Sensitivity to criticism

_____ 40. Fingernail biting or skin picking

ANSWER KEY

Questions 1–10 = Prefrontal cortex symptoms
Questions 11–20 = Cingulate symptoms
Questions 21–30 = Deep limbic system symptoms
Questions 31–40 = Basal ganglia symptoms

If you answered two to three questions related to a particular brain system with a 3 or 4, struggles in that part of the brain may be possible. If you answered four to five questions with a 3 or 4, problems in that brain system are probable. If you answered six or more questions with a 3 or 4, problems in that brain system are highly probable.

Brain Fact #1: You Are Your Brain

As profound as this might sound, the simple fact is that you are a construct of your brain. Let's think about this for a second. Your heartbeat, bodily functions, organs, movements, thoughts, moods, actions, reactions, interactions, personality, memories, health, spirituality, happiness, feelings, relationships, successes, energy, focus, creativity, failures, problem-solving skills, anxieties, diet, decisions, hurts and dreams are all dependent upon the moment-by-moment functioning of the three-pound supercomputer housed within your skull. Your brain is involved in every aspect of your life. It controls everything.

Brain Fact #2: Your Brain Is Ridiculously Complex

As much as we have learned about the brain in just the past decade, we still have not even scratched the surface of understanding how incredibly complicated the human brain is. In fact, many argue that there is nothing in the universe more complicated than the human brain. *Nothing.*

Your brain is estimated to have more than one hundred billion neurons within it, and these neurons have trillions of supporting cells. To complicate things further, each of these trillions of supporting cells can have as many as forty thousand connections (called synapses) between them. This means that a piece of your brain tissue the size of a *grain of sand* has more than one hundred thousand neurons with more than one billion synapses all talking to one another. The critical consensus is that there are more connections in your brain than there are stars in the known universe.

The brain is estimated to hold the equivalent of about six million years' worth of the *Wall Street Journal.* Information travels through your brain at an impressive 268 miles per hour. And although your brain accounts for only about 2 percent of your body's weight, it burns nearly 30 percent of the calories you consume. When we look at the overall temperature of the organs in the human body, the brain is like a massive heat center, burning energy from the food you give it. It works faster and harder than any other organ to manage everything it is responsible for. (This means that you literally are what you eat. Chapter 14 will give you much more insight into the power of food and how it can aid in the healing or hurting of your brain.)

Brain Fact #3: Your Brain Is Not Fully Developed Until Age Twenty-Five

The title of this book was purposely chosen to highlight this critically important brain fact. We may like to assume that we are adults when we turn eighteen; however, the truth is our brain is still undergoing a significant amount of construction until our midtwenties. Research has shown that the brain is not fully developed until a person reaches about the age of twenty-five. For some males, full development can extend until age twenty-eight.

What does this mean for you? If you are under twenty-five, this means every decision you make, every thought you have, every action you take, all the food you eat, the amount of sleep you get and *everything else* you do throughout your day has a significant impact on your developing brain. In short, what you are doing now can affect the rest of your life.

The more effort you make in taking care of your brain and optimizing it now, the better your chances of achieving your goals and dreams. Conversely, the more you harm your brain now, the more difficult things will be in your future. (Chapter 2 will cover the developing brain in much greater detail.)

Brain Fact #4: Your Brain Is Quite Fragile

When we think of the human brain, we often imagine a rubbery, firm organ. This is probably because the brains we have seen outside of skulls are typically kept in formaldehyde, which makes the brain firmer and more rubbery than it really is. Your living brain has about the same consistency as warm butter, an egg white, soft gelatin or soft tofu. If I were able to take the top of your skull off, I could stick my hand in your brain and mush everything around to create one giant mess. Now that's a terrifying thought.

Of course, your brain is surrounded by fluid and housed in a protective skull, but this is often taken for granted. While your skull protects your brain, the brain is still quite vulnerable. When you house something that is very soft in a compartment that is very hard with ridges along the sides that can be as sharp as knives, disaster can strike if proper care is not taken.

I want you to think of some scenarios in which the phenomenon of a soft brain surrounded by liquid and encased in a hard skull would be a negative. Perhaps you thought of Newton's first law of motion: an object in motion stays in motion until acted upon by an external force. In our case, if your head is in motion and then comes to a sudden stop, your brain keeps moving until it smacks against the skull. Then, depending on the severity of the impact, your brain might also bounce the other way and smack against the opposite side of the skull. Anytime this happens, damage to your brain occurs. Car accidents, blunt force trauma and falling out of a second-story window are all obvious ways to damage your brain. What about the less obvious ways we can damage our brain?

Ethan was a bright and popular fifteen-year-old teenager, but his mother was very concerned about him. She was convinced that he was hiding the fact that he was using marijuana or other types of drugs. Over the course of the previous five years, she had seen her sweet, straight-A student transform into a teenager with a 1.3 GPA who was often impulsive, negative and reactive.

The two of them sat across from me in my office. My job was to collect as much information as I could during our two-hour clinical history appointment. I would then give my report, along with his brain scans, over to the psychiatrist for diagnosis and treatment.

As we progressed through Ethan's history, I arrived at the set of questions involving head injuries and concussions. I asked if Ethan played any sports.

"Yeah," Ethan responded in a typical fifteen-year-old's way. "I started playing soccer a few years ago, and I'm pretty good."

"Have you had any concussions from playing soccer?" I asked.

Suddenly, I saw the slightest sense of concern in Ethan's eyes. His mother nearly fell out of her chair. The two made eye contact, but Ethan quickly looked away.

"Ethan had his first concussion when he was eleven years old," his mother said. "It was during his second season of soccer. His coach decided that it would be a great idea for the team to practice headers for forty-five solid minutes. That night Ethan became ill. He complained of a headache and feeling dizzy. I could tell something was wrong, so I took him to the ER. As it turns out, there were six other kids from the team there, too. The coach had given half of the team concussions."

Ethan dropped his head in embarrassment, but I could tell that he wasn't convinced that this could play a role in what was going on with him today.

"Have you had any other concussions?" I asked.

"I've had three more since then," Ethan said.

Hundreds of thousands of children, teenagers and young adults suffer concussions from sports each year. Many of these concussions happen while they are playing soccer, such as when players head soccer balls (hit the ball with their head), a move that has been clocked at seventy-five miles per hour on average. In fact, soccer is the sport that most often results in concussions for female players—female soccer players have a 50 percent chance of getting a concussion while playing. In Ethan's case, his brain scan showed repeated injury, flattening and damage to his prefrontal cortex, the area of the brain in charge of attention, focus, impulsivity, planning and organization. (You will learn much more about this part of the brain in Chapter 3.)

Football is another dangerous sport. Our brains were not designed to be put inside a helmet and then slammed against other helmets. An average tackle on a stationary player can result in up to sixteen hundred pounds of tackling force, with an impact speed of twenty-five miles per hour. Football is the sport with the highest rate of concussions for male athletes. Players have a 75 percent chance of sustaining one.

Nearly 50 percent of athletes who have suffered a concussion do not report feeling any symptoms, which can include headache, fatigue, sleep difficulties, personality changes, sensitivity to light/ noise, dizziness, deficits in short-term memory, difficulty with problem solving, and a general decrease in academic functioning. In some cases, these symptoms are permanent and disabling. And if you have sustained one concussion, you are one to two times more likely to sustain a second one. If you have had two concussions, you are two to four times more likely to sustain a third one. If you have had three concussions, you are three to nine times more likely to sustain a fourth one.

(Chapters 7 and 8 will cover many of the other ways that young people hurt their brains and how this affects their overall chances for a happier, healthier and wealthier life.)

Brain Fact #5: When Your Brain Works Right, You Work Right

Think of what happens when the hard drive on your computer is not functioning at its best. It could be due to a virus, the fragmentation of data, clutter or a host of other issues. What happens when there is a problem with the engine of your car? It doesn't run quite right. Fuel might not be used as efficiently, the engine might be working too hard to get itself going, or it might lose some of the power necessary to move the car effectively.

Many people don't realize that the human brain works in a very similar way. When your brain is working right, you have a much greater chance of working to your full potential. This is when you have the greatest access to yourself—who you really are and what you can really do—and have the ability to achieve the goals you have set for yourself.

But what happens if your brain is not working right? Chances are that you might be having some trouble in your life. This can happen if your brain is underactive, or not working hard enough in certain areas. It can also happen if your brain is overactive, or working *too* hard in certain areas. I'll talk more about this later in the book.

The images you see below show a comparison between two brains that are very different from each other. The brain on the left is an actual brain scan of a healthy person, taken with the use of SPECT imaging (which looks at blood flow and activity). This person's brain is full, even and symmetrical. It's a nice-looking brain. The image on the right, however, is an actual brain scan of a person who has been using methamphetamines for a number of years. You can clearly see global deficits of brain function scattered throughout. It is important to note that this person's brain does not actually have holes in it. The areas appearing to have holes are actually specific regions of the brain that are underactive in terms of blood flow and activity.

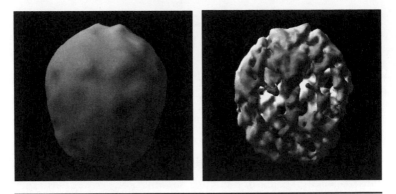

FIGURE 1.1: **HEALTHY BRAIN**   FIGURE 1.2: **BRAIN ON METHAMPHETAMINES**

Throughout this book, you will see many detailed images of the brain that indicate brain injuries and the use of drugs and alcohol. You will also explore the underlying biological mechanisms behind brain struggles, such as attention-deficit/hyperactivity disorder (ADHD), depression, anxiety, obsessive-compulsive disorder (OCD) and many others. (I use the term *brain struggles* to describe a broad spectrum of brain-related issues that people experience, whether it's a diagnosable disorder or simply a behavioral tendency that causes difficulty in a person's daily life.) The important questions you have to ask yourself now as you look at these two images are, Which of these two people has the best chance of success, and which of these two people has the greatest access to his or her true personality and potential?

Brain Fact #6: Certain Parts of the Brain Are Associated with Certain Behaviors

New and exciting research has brought us an impressive amount of information about the brain, though there is an overwhelming amount of information we still do *not* know. One of the more exciting discoveries concerns which parts of the brain are associated with specific behaviors and tendencies. Thanks to the information that has been amassed, we have begun to understand that problems in certain parts of the brain tend to cause specific symptoms and struggles.

Although there are many important parts of the human brain that have been connected to behaviors and success (e.g., the occipital lobes, the parietal lobes, the temporal lobes, the cerebellum and so on), this book focuses on the few parts that relate most to the developing brain.

The figure below gives you a snapshot of some of the areas of the brain that we will cover in great detail in Chapters 3 through 6. Throughout the book I'll explain how many of our strengths, weaknesses and struggles correlate to how specific parts of our brain are working (or not working), and I'll show you how to take practical steps to change the way your brain works.

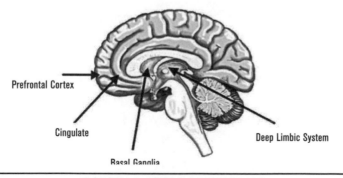

FIGURE 1.3: **AREAS OF THE BRAIN**

Brain Fact #7: Normal Is Not Normal

Over the course of my career as an educator in schools and in the mental health community, I have been continually surprised and disheartened by the level of stigma, shame, guilt, embarrassment and misunderstanding associated with brain struggles. I have learned to ignore the insistence of others that I should just stop being anxious whenever I board a plane. I watched my dad psychologically and physically abuse my sister when she struggled in school, because he thought she was being lazy and needed to try harder, when the truth was she had difficulty with time management, organization and focus because of ADHD. I cringed as my family judged my father for taking his life because of his inability to face his internal demons, while in truth he had serious struggles with ADHD, depression and anxiety.

I have shed tears with students in my office who have struggled with severe depression, even attempted suicide, and have been told by their parents that they just needed to pray harder. My heart has broken for students who have lived in despair and isolation for years, fearing that they were somehow broken or just awful human beings, not worthy of attention or healing. I have seen students blame themselves for their feelings of anxiety and depression, because they didn't know their problems stemmed from a brain that was working against them. In the understanding of the source of those struggles lies hope.

In all the years I have taught others about the brain, the single most destructive obstacle I have faced is people feeling they were not normal because of a brain struggle. The stigma of feeling flawed, less human, isolated or somehow to blame for their struggles is powerful and difficult to overcome.

In the real world, the harsh reality is that normal is a myth. We all have brain struggles. Try to think of a single person who has not struggled from time to time with a focus, attention, organization or time management issue, or with impulsivity, compulsivity, stubbornness, argumentativeness, sadness, irrational worries, low motivation or sleeping difficulties. We all have issues. That is what is normal.

Statistics confirm this reality. Sixty million American adults face some type of brain struggle at any given time. Each year in the United States alone, 5.2 million children struggle with ADHD. And 14.8 million adults battle depression; 40 million, anxiety disorders; 7 million, OCD; 8 million, eating disorders; 5.7 million, bipolar disorder; 30 million, personality disorders; 3.8 million, concussions; 1.7 million, traumatic brain injuries; and 2.4 million, schizophrenia. That's a lot of people dealing with a lot of brain issues. In the end, it is believed that *over half* of the population will be diagnosed with a mental health condition at some point in their lives.

Comorbidity—when a person is facing more than one brain struggle simultaneously—is also a common occurrence. How often do you see people with depression who do not have anxiety issues, as well? People with OCD are often depressed. People with ADHD can also face anxiety and depression. Research suggests that the incidence of comorbidity among people with a brain struggle is as high as 60 percent.

The message here is simple. Our brains are a result of genetics, experiences, traumas, thoughts and the decisions that we make each and every day. To think that it is normal for people to live their life without hardship is naive. Science is confirming this more and more. It's time to put the stigma attached to mental disorders to rest.

Brain Fact #8: You Can Change Your Brain and Change Your Life

When you study the brain and begin to understand its complexity, respect its fragility, appreciate its potential and realize the power you have over how it functions, a spark ignites within. This spark of empowerment and hope can shed light on the darkest of situations. This knowledge can also increase the efficiency and productivity of those who are already doing well. Regardless of who you are or what your situation is, the bottom line is you can be a more successful and joyful person if you embrace the fact that you have the power to change your brain and therefore your life.

There are plenty of ways we can help out our brain. Part III will equip you with a set of strategies and skills to do this. At the same time, we must also pause to recognize the destructive power we can have on our brain. With this knowledge comes a responsibility to avoid the many ways we can harm our brain.

Next Steps

In the next several chapters, you will learn interesting, relevant and practical information about brain development, twenty-first-century changes in the brain, the role the different brain hemispheres play and brain differences based on gender. Then I'll detail four specific parts of the brain that are the source of the most common brain struggles. By the end of Part I, you will probably know more about the brain and how it relates to human behavior than the vast majority of the population. The remainder of the book will tackle (and not in a brain-damaging football kind of way) how you can optimize your brain for healing and a more meaningful life.

SUMMARY

- You are your brain. It controls everything!
- Your brain is ridiculously complex.

- Your brain is not fully developed until about age twenty-five.
- Your brain is quite fragile (think soft butter, an egg white or soft tofu).
- When your brain works right, you work right. When it is in trouble, so are you.
- Certain parts of the brain are associated with certain behaviors. When your brain is having trouble in a particular area, it is usually associated with specific problems, behaviors and symptoms.
- Normal is not normal. We all need some help!
- You can change your brain and change your life.

QUESTIONS

1. How did you feel about your answers to the questions on the Brain Systems Quiz? Is there a particular part of your brain that scored higher than others?

2. How much about the brain did you know before selecting this book to read?

3. What do you hope to gain from reading this book?

THE BRAIN BEFORE 25

Becoming an adult is an interesting concept. Legally, you are an adult the day you turn eighteen. At about this age, you graduate from high school and are supposed to begin a life of your own. You can vote, buy cigarettes, sign binding contracts, join the military, buy lottery tickets, and be arrested and tried in court as an adult.

Historically speaking, the concept that adulthood starts at eighteen has to do with military service and voting. The government decided that boys were mature enough to handle combat at the age of eighteen. Years later, the government decided that since eighteen was the minimum age of eligibility for military service, it should also be the legal voting age. Somehow, this translated into the idea that the brain was fully developed, too, when the individual reached age eighteen.

From a brain science perspective, this is outrageous. As I mentioned earlier, evidence points to the midtwenties as the time when the brain has reached a point where it is functioning at a truly "adultlike" level. For women, the brain is well developed by the age of twenty-five. For men, brain development can extend up to the age of twenty-eight.

One particular industry has known for a long time that people under twenty-five make poor decisions: the car insurance industry. That's why auto insurance rates decrease significantly when an individual reaches age twenty-five. The car insurance industry didn't have to hire an army of brain researchers to figure this out. They had only to look at their own data to find out that most accidents happen when teenagers and young adults are behind the wheel.

The Insurance Institute for Highway Safety reports that teenagers are four times more likely to be involved in a car crash than adults. They are also three times more likely to die in a car accident. In fact, car accidents are the number one cause of death for America's teenage population. This is because teenagers and young adults are more likely to engage in riskier driving, to speed more, to ignore seat belts, to crash while under the influence, to text while driving, to attempt to impress peers while behind the wheel and to underestimate dangerous situations.

Of course, teenagers and young adults often get irritated by facts and statistics like these. I remember feeling this way when I was younger. I heard all the stories and statistics of young people doing dumb things, and I rolled my eyes. I knew that there were dumb teenagers out there, but I also knew that I was not one of them. I was different.

As a young adult, I couldn't figure out why many of the adults in my life were so hell-bent on telling me what I should be doing with my life. I would often hear things like, "Listen to me. I'm older and wiser," "I just don't want you to make the same mistakes I did," "Teenagers think they know everything, but they don't," and "Learn from my experiences." I ignored these pearls of wisdom because I truly believed that I had this "growing up" thing under control. I was smart, I was responsible, and I was eighteen. I was an adult now, right?

Looking back, I cringe at some of the choices I made at that age. Even though my dad was a traditionalist, I thought it was okay to bleach my hair and tell him he couldn't control my life. Yet, I was surprised when all my clothes and belongings ended up on the front lawn in less than ten minutes. I changed majors eight times over the course of seven years in community college because I couldn't figure out what I wanted to do with my life. I also managed to rack up nearly five thousand dollars in credit card debt by the time I was nineteen.

Three days after my twenty-third birthday, my dad was dead. I stepped up and became the executor of his estate, meaning that I was in charge of dealing with funeral arrangements, lawyers, taxes, debt collectors, real estate sales and all the rest of my dad's finances. The estate attorney tried to convince me that I was too young to be handling that level of responsibility. I shrugged off his warnings, but three months later I was crying and overwhelmed by the amount of stress and work that had consumed me.

Ironically, it wasn't until my twenty-fifth birthday that I finally took the reins back in my life. After taking a seat at our dinner table and blowing out the candles on my birthday cake, I sat back and replayed the last seven years of my life. Since the age of eighteen, I had found a certain level of success in life. I was happily married, drove a nice car and lived in a nice home. I was working for Dr. Amen and was enjoying learning about the brain. Still, I wasn't truly happy. I felt ashamed that I had not completed college yet. I felt a yearning to pursue a job that allowed me to work in schools. I hadn't yet started a family of my own.

That night, I made the decision to move forward and pursue my passion of becoming a teacher. I became excited about the possibility of integrating all that I had learned about the brain into the field of education. I told myself that even though seven years had passed since my eighteenth birthday, I would be in a completely different place in seven more years.

And now, seven years later, I have graduated from college and have earned two advanced degrees. I am working as a university professor

continued on page 24

in the education department and am spending time with my wife and two kids, and I just wrote a book about changing your brain and life before the age of twenty-five.

So, what changed between the first seven years after my eighteenth birthday and the seven years following my twenty-fifth birthday? This is where brain science conveniently fits in.

Even though we know that the brain is ridiculously complex and impossible to fully understand, we know a lot about brain development that can be helpful for the teenager and young adult. The following are five brain development facts that will help you understand why teenagers and young adults act and learn the way they do.

Development Fact #1: The Brain Begins Pruning Itself Early

We are born with an overabundance of neurons, gray matter, connections and possibilities. Our brains enter the world ready to take on whatever is thrown at them. Learning, developing and growing happen with amazing speed and intensity.

When we take brain scans of our youngest patients, the heat signatures and activity in their brains light up the screen. Their little supercomputers work much harder than those of adults, and they are like superheated engines of blood flow and activity when compared to their older counterparts. As we age, however, the brain starts to prune itself. It weeds out sections and areas of itself that it has decided it won't use. It begins destroying unused pathways so it can focus on enhancing the ones being used.

Language is a good example of this. When I was growing up, my parents were very proud that I was born an American, so they decided that

I would learn only English. My grandmother, who spoke no English and often babysat me, was barred from speaking to me in Spanish. But I overheard the Spanish-language programs she watched on television and her telephone conversations. That exposure was enough to create the neural pathways in my brain for the Spanish language. Years later, when I took Spanish in high school, not only did I pick it up very quickly, but the intonation and the accent were already ingrained in my brain.

Often, Asians are teased and stereotyped for pronouncing the letter *l* in English as an *r*. I can remember being at my friend's house in high school and watching my friend, who was Chinese-American, trying earnestly to teach his dad the difference between *l* and *r*. After about five minutes, he threw up his hands and had to walk away.

Most of us don't realize that this is a result of brain function. Many Asiatic languages do not make a distinction between the sounds of *l* and *r*. It is not within the language itself. For my friend's father, who was never exposed to English as a child, this meant that his brain had eliminated the neuronal pathways to differentiate between these sounds because he had never used them. As a result, when somebody says "Hello" or "Herro" to him, his brain actually processes it as the same word.

Music is another fascinating example of brain development and pruning at work. With the abundance of gray matter present, children can often learn a musical instrument with ease. Neuronal pathways are created and strengthened. This is different for adults, who have to create new connections and pathways with what they have left in the brain. Can adults learn a new instrument? Of course. Is it harder to do? Absolutely.

Have you ever heard a person with no musical training or experience try to sing a song? As someone who plays piano and therefore has a degree of musical sensitivity, to me it is a sound similar to nails on a chalkboard. Think of *American Idol.* How many times have you seen a person's soul get crushed on national television? The contestants who appear onstage truly believe that they have what it takes to be successful professional singers. When they hear themselves singing, they are unable to detect the inaccuracies in their tone and pitch. This is brain function at work.

So what does this mean for the teenage/young adult brain? It means that the effort you put in now and the amount of new learning you achieve will have a significant impact on your brain in the future. The developing brain is ripe for new information. It is also destroying neuronal pathways, pruning itself to be more efficient at doing what it knows. The goal is to expand what your brain knows today to build capacity for tomorrow.

Development Fact #2: The Brain Develops Back to Front

Until recently, it was assumed that your brain developed all at once. Recent advances in neuroimaging and science have shown us that the brain develops much more specifically, and over a much longer period of time. Scientific studies indicate that the brain develops in a back to front pattern. When we consider this in terms of tendencies and behaviors in childhood and adolescence, it makes complete sense.

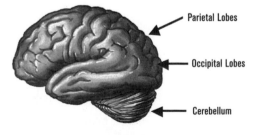

Parietal Lobes

Occipital Lobes

Cerebellum

FIGURE 2.1: **BACK OF THE BRAIN**

One of the first areas of the brain to develop is the cerebellum, which is located at the back of the brain. This part of the brain is critical to human development, as it is in charge of motor movement, sensory information, coordination, balance, equilibrium, muscle tone and much more. And even though the cerebellum comprises only about 10 percent of the human brain, it contains more than 50 percent of the brain's neurons. Think about all the growth and development you see in each of these key areas when people are young. Perhaps this is why young children have such a great desire for physical activity.

As the cerebellum continues to grow, a significant amount of development begins to take place in the occipital lobes, which oversee visual processing and recognition. That's when we see significant development occurring in the parietal lobes, which are involved with cognition, spatial orientation, information processing and speech.

Development Fact #3: The Teenage Brain Is All Drama

Have you noticed that teenagers can be just a tad dramatic and emotional? Oftentimes, we attribute this to hormones and puberty. In doing this, we are not giving the brain the proper credit it deserves.

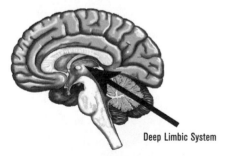

Deep Limbic System

FIGURE 2.2: **THE DEEP LIMBIC SYSTEM**

During the adolescent years, the brain experiences a surge in growth, activity and development in the deep limbic system, which includes the amygdala part of the brain. This part of the brain regulates emotions, feelings and bonding. Of course adolescents are more emotional—it is biologically predetermined.

As adults, we depend on our prefrontal cortex to regulate and supervise the rash and emotional behaviors stemming from our deep limbic system. As you will learn later, the prefrontal cortex is what facilitates logic, insight and an understanding of past experiences, as it filters out the nonsense that often spews out of our emotions. The problem is that teenagers do not have a developed prefrontal cortex to supervise the onslaught of emotions and drama that comes with high school. When they *feel* strongly about something, they will often make a decision based on their emotions rather than logic.

Do you remember your first love? Or your first kiss? Do you remember how powerful that attraction was, and how it consumed your thoughts, time and emotions? Many adults look back at their first love with great fondness. I've heard some adults say the love they felt for that person as a teenager was stronger than anything else they have experienced—even as adults. That's because your brain is in its strongest emotional state during this period of life, and there is not much controlling or filtering of those emotions.

Think of the amount of time that teenagers invest in their social lives. For many, high school is all about friendships, being popular and receiving positive feedback from peers. Essentially, the brain is operating from the amygdala while feeding the reward centers. And even though logic might suggest that teenagers' time would be put to better use studying and preparing for the future, the feelings and emotions are just too strong.

Development Fact #4:
Peer Pressure Doesn't Come from Peers

I have had countless conversations with frustrated parents who are surprised by the actions of their adolescent children. Often, peer pressure becomes the focus of the discussion. "It was that hooligan friend who led my child astray." "It had to be the influence of somebody else, because I know I taught my child better than that." Unfortunately, this focus on peer pressure is more often a cop-out than a true explanation of what is happening.

Think about all the peer pressure campaigns you were exposed to in school. They all have the same story line. You are hanging out with friends one day, having a good time. Then, out of nowhere, one of your friends decides that it would be "cool" to _____. You can fill in that blank with a number of different scenarios: smoke a cigarette, drink alcohol, tease another student, try drugs and so on. Next, you are supposed to step back and think about it. You know it's wrong. You know you don't really want to do it. The problem is that you are being pressured into doing it. You were taught to "just say no." You are supposed to stand up for yourself and do what's right.

From a brain science perspective, this is utter nonsense. The concept of peer pressure removes responsibility from the individual and assumes the pressure comes only from external forces. This is comforting on the one hand, because it means there is someone else to blame. The reality of the situation is that the vast majority of adolescents who make poor decisions actually do so on their own. This is where things get interesting.

Already, you have learned that the adolescent brain is unique in its development and activity level. It is going through its days without much help from the prefrontal cortex. Decisions are largely based on emotions from a developing and overactive deep limbic system. At the same time, there is another part of the brain that is highly activated during the adolescent years: the reward centers.

In adolescence, neuronal connectivity and development in the reward centers of the brain reach their peak. This means that teenagers will often do things that give them immediate pleasure and/or gratification. These things can be related to alcohol, drugs, sex and social praise. And, as you already know, teenagers are still lacking that critical part of the brain (i.e., the prefrontal cortex) that serves to keep things in check. This means that teenagers will often *choose* on their own to do certain things because of the rewards they seek. Oftentimes, the reward they seek is praise and recognition from their peers.

In 2010 researchers at Temple University conducted a study to look at the differences between the teenage and adult brain in terms of peer pressure and the reward centers of the brain. In the study, teenagers and adults were asked to play a driving video game while having an MRI scan done. This allowed researchers to see the functioning of the brain as the subjects were playing the video game. Initially, there was not much difference in the data that was collected from teenage and adult brains.

Participants were then told that a group of friends was watching them play the video game from another room. When this happened, the reward centers of the brains of the teenagers became activated, and the teenagers ran 40 percent more yellow lights and had 60 percent more accidents during their video game play than their adult counterparts. Nobody in particular was pressuring the teenagers to engage in riskier behaviors. None of their peers were even in the same room. It was the mere suggestion that their peers were observing them that changed the behavior of the teenagers.

This is an important finding as it illustrates not only that peer pressure definitely exists, but also that we may not recognize that it is much more internal than it is external. In the Temple University study no one convinced or pressured the teenagers to act out; the teens acted out because they wanted to impress their friends. Teenagers might choose to engage in risky behaviors because their need for social acceptance, recognition and friendships overpowers them. Plus, they do not have full access to the part of the brain that regulates everything: the prefrontal cortex.

Development Fact #5: The Prefrontal Cortex Is Last to Develop

You've been hearing a lot about the prefrontal cortex, and you will learn more about it in Chapter 3. For now, just remember that the prefrontal cortex is the regulator of our lives. It stops us from making poor decisions, and it helps us plan, stay focused and think logically. It allows us to look at the different sides of a situation and reminds us of the consequences of our actions. It is a pretty remarkable and important part of who we are. Unfortunately, this critical component of our brain is the last part to develop fully.

What does this mean for the teenage and young adult brain? It means that we have to recognize we are not quite the masters of our domains during this time.

It also means that we cannot blame others for our mistakes and failures. By reading this book, you are empowering yourself with an incredible wealth of knowledge about the human brain. You will be given a set of skills and strategies that will enable you to change your brain and your life. With dedication, you have the chance to lay the foundation for a healthier, wealthier and more joyful life. Think of it this way: every decision you make throughout your day has the power to change your brain for better or worse.

SUMMARY

- The brain begins pruning itself early. This means that you have the power to construct some incredible connections and pathways in your brain by learning while you are young.

- The brain develops from back to front. It starts with physical movements and coordination and ends with focus and forethought.

- The teenage brain is all drama. The deep limbic part of the brain, which is all about emotions, relationships and bonding, is highly active.

- Peer pressure is much more internal than external.

- The prefrontal cortex is the last part of your brain to develop fully. This is the part of the brain responsible for attention, focus, concentration, organization, procrastination, forethought, judgment and planning.

QUESTIONS

1. Was there ever a time when you felt like the limbic part of your brain (emotions) got in the way of your judgment?

2. Was there ever a time when you felt more pressure from yourself than from your peers?

3. What are some things that you would like to begin learning to help your brain develop pathways and connections now?

3 ANTICS AND THE PREFRONTAL CORTEX

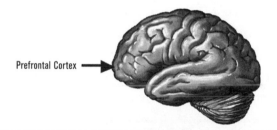

FIGURE 3.1: **THE PREFRONTAL CORTEX**

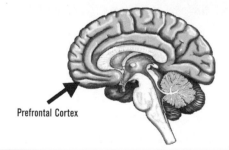

FIGURE 3.2: **THE PREFRONTAL CORTEX (2ND VIEW)**

The prefrontal cortex is one of the largest structures of your brain, taking up the front third of it. Behaviorally, it plays an incredibly critical part in who you are. Think of it as the CEO of your brain and your life. It takes in all your thoughts and inputs from your senses and then orchestrates new thoughts and coordinates with the rest of your brain to achieve specific goals. It is the part of your brain that allows you to focus, pay attention and stay organized. It allows you to plan, learn from mistakes and use judgment when thinking of what to say and do. It is also involved in forethought, impulse control and follow-through.

Many experts argue that the prefrontal cortex is what makes us human. It acts as our conscience, and it keeps us from doing dumb things. In a way, the prefrontal cortex is the Jiminy Cricket of our lives. It's that little voice that reminds us what is right and wrong. It helps us control our actions, consider the consequences and plan accordingly.

The prefrontal cortex is much more prominent and advanced in the human brain than in the brains of other animals. Whereas the prefrontal cortex is about 30 percent of our brain, it is only about 17 percent of the chimpanzee brain, 11 percent of the gibbon (a type of ape) brain and 7 percent of the dog brain. If you are a cat lover, you might want to skip the next few sentences, because, sadly, the prefrontal cortex is only about 3 percent of the cat's brain. Think of how much control a cat has over its actions. How capable is a cat of planning, focusing, sustaining attention, learning from mistakes or inhibiting impulses? I have two cats at home, so I can answer this with two words: laser pointer.

For a student in high school, understanding the prefrontal cortex is critical because it is still undergoing tremendous development. Have you ever stopped to measure the amount of focus that a typical high school student has in any given class? Unless he or she has a highly engaging teacher or is particularly connected to the content, the answer is probably "Not much."

Prefrontal Cortex Functions

Forethought is the ability to stop and consider the consequences that might result in the future because of your actions now. Planning is similar in that it helps you achieve the goals that you have laid out for yourself. And, of course, judgment is the process of considering decisions or coming to sensible conclusions after careful thought. In a developing teenage brain, these traits are a work in progress.

I was recently with my son at a local amusement park for his fourth birthday. As we were walking through the children's area of the park, I noticed two high school students skipping and laughing wildly. The teenagers were eager to get on as many children's rides as possible, and I knew from personal experience that they were sure to make fools of themselves in the name of fun. After my four-year-old got in line at the Wild Frog Ride, I cringed when the two teenagers joined the line behind us. As they joked with each other, making loud noises and pushing each other in and out of line, I discreetly bribed my son with ice cream to get out of line and go somewhere else. I remembered being that age and doing the same thing, so I knew this was going to end badly.

As we waited in line for the mini Ferris wheel, I watched as the two teenagers sat on the Wild Frog Ride with five kids ranging in age from about three to six years old next to them. When the ride started, the teenagers began screaming, acting as though they were terrified out of their minds and about to die.

Adults do not act out in this way owing to the fact that forethought and judgment typically step in. If adults ever found themselves contemplating such behavior, they would ask themselves, *Is it a good idea to scream bloody murder on a children's ride when you are seated next to numerous children who are barely in school? What would be the effect of this? And is it necessary?*

As the teenagers staged their impromptu dramatization of a horror film on the kiddie ride, I watched as the blood drained from the neighboring children's faces. Within seconds, all the kids on the ride were crying uncontrollably.

The two teens on the ride saw the kids crying and began laughing about how they had scared them. I could see parents pacing back and forth on the ground nearby, no doubt pondering the therapy bills they would need to start saving for as a result of this trauma. Kids who had been waiting in line to go on the ride next were instead running to be comforted by their parents and to get as far away from the ride as possible. I could also see a couple of dads gearing up to give the teenagers an earful as soon as they exited the ride.

At school, forethought and judgment help us to realize that paying attention in class and getting good grades are in our best interest for success in life. They help us understand that arguing and behaving rudely with our teachers usually does more harm than good. Planning is important because it helps us realize that the ten-page final paper is something that we should start a few weeks in advance and not the night before it is due.

Being impulsive means acting without forethought—in other words, converting your thoughts into words or actions without stopping to consider the consequences. Empathy involves the ability to understand and experience the feelings of another. Learning from your mistakes is fairly self-explanatory. It means that you make a mistake, learn from it, and then stop and think about the past before you repeat the same mistake again. These are all functions of the prefrontal cortex.

I remember when I was in high school and we had a substitute teacher in our class for about a week. The sub was an older woman who was soft-spoken and not very good at containing the class or relating to the students. This, of course, meant that the class became a free-for-all. Many of the more impulsive students spent their time engaged in full conversation as the sub tried to teach from the front. Whenever the sub had her back to the class, some of the students would throw spit wads and balled-up pieces of paper.

On the fourth day, two of my classmates began making jokes about cancer. The substitute teacher stopped teaching and sat down at the desk. We could see the tears welling up in her eyes. A few of the girls in the class started to yell at the boys to shut up. They obviously saw that the conversation was having a profound impact on the sub's feelings. Instead of quieting down, one of the boys replied, "Who cares? It's not like she has cancer or something. We're just messing around."

The sub looked up and glared at the boy who had made the comment. Slowly, she raised her hand to her head and removed the wig she had been wearing. Underneath the wig was her balding scalp, a side effect of the radiation treatment she had been receiving for breast cancer. Then she stood up and walked out of the class.

Problems with the Prefrontal Cortex

If you are under the age of twenty-five, the cruel reality is that you are going to struggle with many of the functions of the prefrontal cortex simply because it is not yet fully developed. This means that you will need to work harder to ensure that you are able to focus, exercise forethought, control impulses, organize yourself, plan appropriately, set goals, show good judgment, have empathy for others, control your emotions, develop insight and learn from your mistakes.

For many teenagers and young adults, however, problems with the prefrontal cortex run deeper. Oftentimes, this is due to this region of the brain not working as hard as it needs to. The image below illustrates what the prefrontal cortex of a young person's brain looks like when it is underactive. Note the two "holes" in this area of the brain. Remember, these are not actual holes in the brain; instead, they indicate that the prefrontal cortex of this person's brain is underactive compared to the rest of the brain.

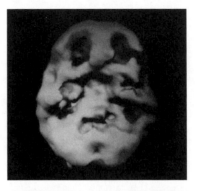

FIGURE 3.3: **UNDERACTIVE PREFRONTAL CORTEX**

When a person has problems with his or her prefrontal cortex, it is almost like the supervisor is missing. There might be pronounced difficulty with attention span, a desire to live in the moment, impulsivity, disorganization, procrastination, poor judgment, a lack of empathy, a failure to pay close attention to details, a lack of insight, trouble learning from mistakes, trouble listening, a tendency to lose things, a tendency to be easily distracted, poor planning skills and a lack of clear goals or forward thinking.

It is important to note that we can all relate to behaviors like these to a certain degree. We all have our moments of disorganization and procrastination. However, when these behaviors become the norm, they impede your joy and your ability to succeed in school and in life. The person with prefrontal cortex issues is disorganized to a high degree. Others will look at this person's bedroom, locker or backpack and wonder how someone can live that way. But to the person with prefrontal cortex issues, it is simply life. In fact, this person might even argue that even though his or her room or locker or backpack appears to be disorganized, he or she knows where *everything* is.

This person will also procrastinate regularly, often waiting to start an assignment until the night before it's due, the morning of the day it's due or even the day after it's due. This person might also be impulsive and lack empathy. I know a couple that was having difficulty because the husband had ADD, which made him prone to saying whatever came to mind. Early in their marriage, the wife put on a new yellow dress for a night out. When she asked her husband his opinion of the dress, he blurted, "You look like a school bus." As the wife backed away in horror at the insult, her husband made the beeping sound of a bus backing up. He thought it was hilarious and had no idea how hurt his wife was by his remarks. If confronted about their behavior, people like this man will argue that they are just being "real" and telling it like it is.

Conditions Associated with the Prefrontal Cortex

The condition that is most commonly associated with the prefrontal cortex is ADD/ADHD. Some of the symptoms of ADD/ADHD include failing to pay close attention to details, difficulty sustaining attention, failing to listen when spoken to, failing to follow through on instructions, difficulty organizing tasks, a tendency to avoid tasks that require sustained mental effort, a tendency to lose things, a tendency to become easily distracted and forgetfulness. The hyperactive component of ADD/ADHD includes such symptoms as fidgetiness, difficulty remaining seated, running/climbing excessively, restlessness, difficulty being quiet, talking excessively and a tendency to be often "on the go." Lastly, the impulsive component of ADD/ADHD includes such symptoms as difficulty waiting your turn in games or group situations, blurting out answers and interrupting or intruding on others. In order for an individual to be officially diagnosed, symptoms have to cause impairment at school, at work or in life, and must be present for at least six months.

Problems with the prefrontal cortex have also been associated with other types of mental disorders, including brain trauma, schizophrenia, conduct disorders, some types of depression, dementia and antisocial personality disorder.

When I was an elementary school teacher, I often encountered children who lacked judgment, forethought, planning and organization, but one particular student will always stand out for me when I think about problems with the prefrontal cortex.

Garrett was the class clown. He loved getting attention, even though most of it was negative. His desk was a mess, his backpack boasted remnants of his lunches from the previous three weeks, he couldn't tie his shoes to save his life and he received an abnormal amount of joy from irritating as many of his classmates as possible. Oh, and this was the fourth grade.

Over time, and with hard work from me as the classroom teacher, his peers learned to ignore Garrett's behaviors. As expected, he became frustrated that his typical antics were falling on deaf ears. Determined to get the reaction and attention he craved, he started to push his boundaries even further. The culmination of this determination was an event that is forever etched into the communal memory of our school.

It happened after a school assembly one crisp February morning. Garrett had been complaining that his stomach hurt. This was not a surprise since he often complained of something in hopes of getting to go home early. When no one responded to him, he started to groan louder and complain with more intent. After a few minutes of his escalating attempts at faking sickness, I called him to the front of the line and told him that he could visit the nurse during recess. I have found most students who are not feeling well magically begin to feel better when they are told they will miss their recess time. This news did not sit well with Garrett, because he knew he wasn't going to get his way. He stared me down with the angriest look he could muster, and then I watched in disbelief as he raised his index finger in the air, opened his mouth and then proceeded to make himself gag...never breaking eye contact with me. A quick note: Throwing up in your own hands is a surefire way to get attention. Lots of it. Very quickly.

As I watched Garrett fill his cupped hands with vomit, I was horrified and unable to react. But the students' responses were immediate. Some children began screaming. Others cried out in horror and leaned on one another for comfort. Some students started dry heaving, trying desperately to hold in their own vomit. What happened next became the single most memorable moment of my teaching career: Satisfied that all the attention was on him, Garrett smiled. Then, with all eyes on him, he looked at me, winked, raised his hands to his mouth and started to drink his own vomit.

To be honest, I am unable to recall many of the details of what occurred after that moment. I vaguely remember feeling like I was in the midst of a Hollywood disaster movie. Children were screaming and running in different directions. I am fairly confident that I gently pushed some of the children out of my way during my sprint to the door, desperate to get outside before I vomited all over the place, too.

Now Garrett is an extreme case, one that is not the result of a developing prefrontal cortex but rather a damaged one. Both of his parents suffered from drug addiction, so Garrett was raised by his grandmother. She later told me about an incident when Garrett was just three or four years old. He fell down a few stairs and landed on his forehead—exactly where the prefrontal cortex is situated. The doctors in the emergency room thought he experienced nothing more severe than a mild concussion, but Garrett's grandmother said that after the accident, the boy was never the same.

What Does It All Mean?

The prefrontal cortex is perhaps the most critical part of the human brain because of all it is involved with. Without the appropriate ability to plan, organize, focus, set goals, use judgment and feel empathy for others, life is fairly difficult to navigate.

It is important to remember that this part of the brain is still under rapid development in the teenager and young adult. Too often, I have witnessed teachers and parents spending too much time disciplining and judging their children or students, assuming that they are just not trying hard enough or are not applying themselves. This means that we adults need to ensure we have the proper supports in place to enable young people to make the right decisions.

On the flip side, this doesn't mean that teenagers and young adults get to make excuses for their actions. If anything, the knowledge you have accrued thus far about the brain should encourage you to work harder to ensure your own success. Over the years, I have taught countless high school and college-age students about the brain. I have always been intrigued to see the reactions of young people as they learn about the human brain and how it relates to behavior and success in life. When we cover the prefrontal cortex, I typically see two types of responses from students:

- **Response one:** Young people who squeal with excitement and then immediately devise a rationalization of their previous actions and behaviors. I have even had students tell me that they were unable to submit their homework assignments to me because they have an underactive prefrontal cortex, and that I needed to give them extra time.

- **Response two:** Young people who are energized by the information and take the necessary steps to ensure they make up for any underdevelopment or underactivity of their brain. These young people are not looking for a crutch or an excuse. Instead, they attempt to compensate for any perceived shortcomings and work harder to ensure they achieve the level of success they desire.

In the end, however, my response to all these reactions is the same. Regardless of who you are and what brain you were born with, you alone have the power to encourage the growth and development of your prefrontal cortex in a way that will foster new learning and success in your life. You can set up supports to help compensate for the lack of maturity within this critical region of the brain. This means reflecting on the comments, suggestions and advice given to you by older people, such as parents, teachers and other mentors. Usually, these individuals are truly trying to help young people avoid the mistakes they made in their own past. Now that they are older, wiser, and have a brain and a prefrontal cortex that is more fully developed, they are able to think more logically and realistically, and with greater forethought and judgment. They are, in effect, trying to be the prefrontal cortex for you. They may not be right all the time, but it might be wise to step back and allow yourself to seriously consider what they might be saying.

SUMMARY	
PFC Functions (supervision)	**Low PFC Problems (lack of supervision)**
» Focus	» Short attention span
» Forethought	» Lack of clear goals or forward thinking
» Impulse control	» Impulsivity

continued on page 44

PFC Functions (supervision)	Low PFC Problems (lack of supervision)
» Organization	» Disorganization
» Planning, goal-setting	» Procrastination
» Empathy	» Lack of empathy
» Emotional control	» Failure to pay close attention to detail
» Insight	» Lack of insight
» Learning from mistakes	» Trouble learning from mistakes
	» Tendency to lose things
	» Tendency to be easily distracted

Having low PFC activity may entail some positive traits, such as:	
» Spontaneous	» Creative
» Not rule bound	» Uninhibited, free-spirited
» Potential to be a great salesperson (as long as you have an assistant who keeps you organized)	

Conditions associated with low PFC activity include:	
» ADHD	» Some forms of depression
» Brain trauma dementia	» Bad judgment
» Schizophrenia	» Antisocial personality disorder
» Conduct disorders	

Common treatments implemented when PFC activity is low:	
» Organizational help	» Intense aerobic exercise
» Goal setting/ planning exercises	» Stimulating or exciting activities
» Relationship counseling	» Neurofeedback
» Higher-protein diet	» Working on developing a deep sense of personal meaning

QUESTIONS

1. How do you think your prefrontal cortex is operating?

2. What can you relate to in this chapter? Does the information in this chapter shed light on the behavior of someone in your life?

3. What are some things you might do differently in your own life now that you know about the prefrontal cortex and what it controls?

4 THE CINGULATE GYRUS AND COGNITIVE FLEXIBILITY

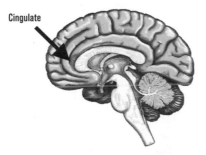

Cingulate

FIGURE 4.1: **THE CINGULATE**

If the prefrontal cortex is the central computer of a car, then the cingulated gyrus—what Dr. Amen and I commonly refer to as simply "the cingulate"—is the transmission, or gearshift, the part of the car that allows the engine to shift from gear to gear and ensures that it operates smoothly. The cingulate portion of our brain is what allows us to shift from thought to thought, from action to action, from behavior to behavior and from task to task. To get a better understanding of the cingulate's position inside the brain, imagine you have a Mohawk.

Now run your finger along it, down the center of your scalp from your forehead to the back of your head.

Cingulate Functions

We call the cingulate the brain's gearshift because it allows us to shift our attention, be flexible and learn to adapt to change when needed. It allows a person to go with the flow, see options and let go of negative thoughts.

In school, the cingulate is what allows us to close the book at the end of one class and move on to the next class and focus on a new subject. When you receive feedback on an assignment, the cingulate is what allows you to interpret the feedback as valuable and helpful—for instance, as enabling you to write a stronger paper and achieve a higher grade in the class—rather than as criticism. A healthy cingulate also allows you to focus on the big picture, and it keeps you from obsessing about the small details and wasting energy trying to be perfect.

In friendships and relationships, the cingulate is important because it enables us to accept others' flaws rather than trying to change them. When a person with an underactive prefrontal cortex says something impulsive and immature, our cingulate allows us to forgive, let go and move on.

The cingulate also allows us to remind ourselves that being argumentative with people, particularly our parents, really does not serve our best interest.

Problems with the Cingulate

Like the prefrontal cortex, the cingulate can also have difficulty. The images below show what a struggling cingulate looks like. In these pictures, the areas of white are hot spots of overactivity—areas that are working much harder than the rest of the brain.

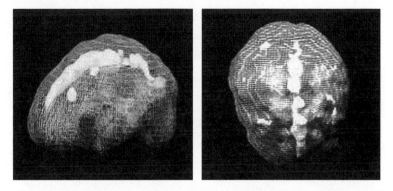

FIGURE 4.2: **OVERACTIVE CINGULATE, SIDE VIEW**

FIGURE 4.3: **OVERACTIVE CINGULATE, TOP VIEW**

When the cingulate is working too hard, your brain may have a tendency to get stuck. Instead of being able to shift from thought to thought and from action to action, your brain may become obsessive and compulsive.

I can always spot the overactive cingulates in the classroom. If I switch activities, those with hot cingulates will be the ones who cannot shift their attention. They *must* finish the work in front of them before they are able to move on. They might get stuck on one particular problem. I have even seen some students hide their homework worksheets from a previous topic inside a textbook being used for the current activity because they "just had to finish."

People with a struggling cingulate are often labeled oppositional, defiant, stubborn or argumentative. They believe that they are always right, and they might spend an exorbitant amount of time trying to show others *why* they are right. Often they argue just for the sake of arguing, and they have an automatic "no" response. They worry excessively and become fixated on certain thoughts, feelings or actions.

If you offend a person with an overactive cingulate, he or she will beat you over the head about all the wrongs you have ever done. You'll find this person can never seem to let it go. This person can hold grudges for a lifetime and may replay events in his or her head to devise better ways to have handled previous situations. Take the woman in Chapter 3 whose husband said she looked like a school bus in her new yellow dress. She knew full well that her husband suffered from ADD, and that his comments were likely a result of that. But because she herself had struggles with her cingulate, twenty-five years later she was still talking about her husband's insult. Rather than chalking his insult up to his brain, her cingulate couldn't get past it.

My cousin Gilbert is a bright and talented guy, with a particular passion for heading soccer balls competitively. In school, most of his classes came easy to him, and he excelled in athletics and making friends. His downside, however, is that he has a hot cingulate, which means that he can be incredibly stubborn.

In high school, his grades suffered because he often felt he didn't need to listen to the teachers. He got in trouble regularly for making snide remarks in class, and he had a particular knack for getting into arguments with his teachers, parents and other authority figures. His attitude was, "I know what I'm doing, and I don't need people telling me what I should be doing instead."

Gilbert was surprised when he was rejected from every university he applied to. I remember chuckling to myself when he placed the blame on his teachers, coaches and parents for not enabling him to earn the GPA he "truly deserved."

During his first year at community college, he continued the same pattern of stubbornness and opposition. He also became obsessed with the video game *Call of Duty,* which he played on his PlayStation. Soon after, his parents began to grow tired of his attitude and disrespect.

The following is his story, in his own words, of how things erupted into a family war.

That Sunday morning started off like every other Sunday. We woke up early, went to mass and then came back home to relax for a bit. Later that day, a couple of my younger cousins came over to my house to spend some time with us.

As much as I love my little cousins, they kind of irritate me, because all they want to do is play my video games and break things. That day, while I was in the middle of an epic Call of Duty *match, they were asking me if I could set up another video game console for them in the living room.*

After being rejected by me a few times, they decided to go ask my mom, because they knew she would say yes. Of course, my mom then came into my room and yelled at me to set up the game for them.

I told my mom in a disrespectful way that she needed to hold on so I could finish my game. My mom marched over to me and shut off my game. As I stomped into the living room to set up the game for my little cousins, I made sure to roll my eyes and give as much attitude as I could.

My mother became even more upset. She told me how ungrateful I was being, and how she could easily shut off the Internet and take away my video games so that I would have nothing to play.

"Go ahead!" I said. "I can just go ahead and buy another game and buy my own Internet subscription. What makes you think I need you to buy stuff for me!"

At this point, my mother lost all her cool and started screaming about how the clothes I was wearing were clothes that she had bought for me, and how I needed her to survive.

continued on page 52

I became even more upset and told her I didn't need anything from her. Then, in front of my two little cousins, I took off my jeans and shirt and threw them on the floor next to my mom. And then, wearing nothing but my boxer briefs, I marched into my room and slammed my door shut.

My dad, who had been in the backyard and had been listening the entire time, went nuts and came into my room to start yelling at me. I screamed back at him in complete defiance, which is something I had never done before. My dad struck me on my chest and then kicked me out of the house.

Gilbert's cingulate, combined with the cingulates of his parents, resulted in a situation that spiraled quickly out of control. What should have been a joyful day with family turned into a day filled with tears, fear and actions that everyone immediately regretted.

Coincidentally, I arrived in town late that night to stay at their house for a few days while attending a conference close by. I could tell right away that something was wrong. My aunt's eyes were swollen and red from weeping, and Gilbert was nowhere to be found. After she filled me in and we tracked Gilbert down at a friend's house around 1:00 a.m., I went to pick him up. Then we drove to Denny's.

For the next two hours, Gilbert and I talked about the human brain. I told him that the brain of an eighteen-year-old was not as mature and developed as we might think it is. We talked about the prefrontal cortex and how forethought and judgment are important to our lives. And we spent a great deal of time talking about the cingulate. I was able to show him how his brain was wired in a way that made him stubborn, argumentative and defiant by nature. More important, I explained to him that our brains are the way they are in part because of genetics. If he was stubborn and always wanted to be right, it was safe to assume that his parents were the same way. While his cingulate was running at full speed that day, he triggered the cingulates of his parents, turning the entire situation into one massive war of stubbornness and argumentativeness.

I made sure that Gilbert understood that I wasn't there to tell him he was wrong. I wasn't there to yell at him or make him feel bad for what had transpired that day. I only wanted him to understand *why* things might have happened the way they did. I also tried to help him understand that it was in his best interest to keep his cingulate at bay when dealing with his parents. I did this by asking him a few simple questions, including:

- How many times have you argued with your parents and changed their viewpoint?

- When has your negative attitude toward your parents ever resulted in anything positive for you?

- When you act stubborn, argumentative, oppositional or defiant with your parents, what typically happens afterward?

- How would your parents engage with you if you never argued with them? Would anything change? Would they be happier? Would you be happier?

In answering these questions, Gilbert began to see the futility of his actions. The harsh reality of the situation was that Gilbert truly was dependent on his parents. Without them, he would not have a home, a car, insurance, food or the Internet.

I'll talk more about what happened in Chapter 11: Adapting Your Brain to the Brains of Others.

Conditions Associated with the Cingulate

When the cingulate is overactive, many things can go wrong. An overactive cingulate can manifest itself as a personality that is stubborn, argumentative, oppositional, defiant, and even somewhat obsessive and compulsive.

For others, however, an overactive cingulate can lead to conditions that are debilitating and even devastating. Obsessive-compulsive disorder, for example, is a condition in which a person *must* perform certain meaningless actions repeatedly in an attempt to alleviate obsessive and irrational fears or intrusive thoughts. The cingulate is unable to shift gears, and this person will get stuck on the same thought, image, impulse or worry over and over again. As much as this person might want to stop this behavior, he or she is unable to do so.

Examples of compulsions that arise with OCD include checking, washing, counting, arranging and hoarding. For the most part, these compulsions are done in an attempt to make the obsessions go away. Sadly, relief from the obsession through the act of the compulsion never lasts, and the obsession will typically come back stronger than before.

Eating disorders are also associated with the cingulate, as they are marked by the inability to let go of the thought of being overweight. Even though all the people in your life might tell you that you're too skinny and unhealthy, a person suffering from an eating disorder cannot escape obsessive thoughts about weight. If the transmission is stuck, the brain cannot shift gears.

Chronic pain, some anxiety disorders, certain phobias, oppositional defiant disorder, road rage and even addictions can also be associated with the cingulate. From a brain science perspective, this makes sense since they all relate to the inability to shift away from a particular thought or action. The cingulate needs to calm down in order for healing to occur.

What Does It All Mean?

It is important to nurture cognitive flexibility in a teenager or young adult who is trying to navigate through school, a new career and relationships. Being flexible and open to ideas, and possessing the ability to let go of unnecessary worries and hurts, helps us to grow, develop and mature in all areas of life.

We need the cingulate to be *active* in order to be successful. People with particularly active cingulates can be very focused, goal oriented, organized, on track and predictable, and they are often good with numbers. But it is important to recognize when a cingulate is *overactive* and is doing more damage than good. We can all relate to having an overactive cingulate at times. But if you find yourself repeatedly unable to let go of troublesome thoughts, or if you find yourself obsessing about thoughts or engaging in actions that are impeding your joy, then you know it might be time to seek help.

However, telling a person with an overactive cingulate that he or she has an overactive cingulate never goes well. The immediate response is typically, "No, it's not." If you recall, this was my reaction when my own father-in-law pointed out my stubbornness, which, I later learned, he had detected from my overactive cingulate in my own brain scan.

What is important to remember is that these tendencies all stem from brain function. If you can relate to much of the information in this chapter, if you see certain attributes delineated in this chapter in yourself or in others in your life, there is a lot you can do about it. Chapter 10 and all of Part III provide you with plenty of tips and strategies to calm down the cingulate and pave the way for a calmer and brighter tomorrow.

SUMMARY	
Cingulate Functions	**Cingulate Problems**
» Brain's gearshift working efficiently	» Getting stuck on negative thoughts or behaviors
» Cognitive flexibility	» Worries
» Cooperation	» Holding grudges
» Moving from idea to idea	» Obsessions/compulsions
» Seeing options	» Being inflexible, may appear selfish
» Going with the flow	» Oppositional/ argumentative
	» Feeling upset when things do not go your way
	» Feeling upset when things are out of place
	» Having an intense dislike of change
	» Tending to say no without thinking

Having high cingulate activity may entail some positive traits, such as:	
» Very focused	» Staying on track
» Goal-oriented	» Predictable
» Very organized	» Has the potential to be a great accountant or number cruncher

Some medical conditions associated with high cingulate activity are:

» Obsessive-compulsive disorder	» Post-traumatic stress disorder
» PMS	» Eating disorders, such as compulsive overeating
» Chronic pain	» Addictions
» Oppositional defiant disorder	» Tourette's syndrome
» Difficult temperament	

Common treatments implemented when cingulate activity is high:

» Distraction, creating options	» Intense aerobic exercise
» Neurofeedback	» Relationship counseling
» Anger management	» Lower protein/ complex carb diet

QUESTIONS

1. How do you think your cingulate is operating?

2. What can you relate to in this chapter? Does the information in this chapter shed light on the behavior of someone in your life?

3. What are some things you might do differently in your own life now that you know about the cingulate and what it controls?

5 EMOTIONS AND BONDING: MEET THE LIMBIC BRAIN

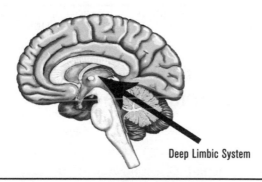

Deep Limbic System

FIGURE 5.1: **THE DEEP LIMBIC SYSTEM**

As you read in Chapter 2, the deep limbic part of the brain becomes especially active during the teenage and young adult years. This surge in activity often causes quite a bit of drama and chaos in our lives, relationships and friendships.

The following images show a comparison of two brains. The scan on the left illustrates what a healthy brain scan looks like. Notice that the only areas of white are at the bottom of the brain, in the cerebellum. This is to be expected, as more than half of the brain's neurons are located in the cerebellum.

The scan on the right indicates that several areas of the brain are working too hard. The arrow points to the overactive deep limbic system. Due to this overactivity, this person was struggling with depression, low motivation and thoughts of suicide.

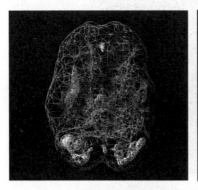 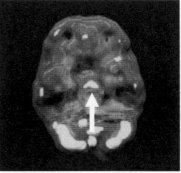

FIGURE 5.2: **HEALTHY DEEP LIMBIC SYSTEM**   FIGURE 5.3: **OVERACTIVE DEEP LIMBIC SYSTEM**

Functions of the Deep Limbic System

Like the prefrontal cortex and cingulate parts of the brain, the deep limbic system is a critically important part of our brain and our lives. To illustrate all the different functions of the deep limbic system, let's look at a typical day in your life.

First and foremost, the amount and the quality of sleep that you get is dependent on the functioning of your deep limbic system. This means that the deep limbic system determines how tired, energetic or lethargic you might feel when you wake up in the morning.

Your limbic system also controls your motivation and drive. This means that after you wake up, your limbic system is what actually gets you out of bed to start the day. It helps you recognize that lying in bed all day and wasting time may not be in your best interest. It gets you excited about the events of the day, the people you plan on interacting with, and it drives you to get things moving.

Of course, some days, you just feel that it is going to be a bad day. Other days, you wake up feeling sprightly and energetic. Your emotional tone and mood are centered within your deep limbic system. It sets the tone and then acts like an emotional filter for whatever happens on the outside.

Have you ever been around a Debbie Downer or a Negative Nancy? These are people who are negative, pessimistic and a bummer to be around. I remember working with a teacher who was this way. Mrs. James would drag her feet into the faculty office each morning and complain about anything and everything. If somebody commented on the nice, sunny weather, Mrs. James complained that the sun made her break out in hives. If someone observed that it was nice to have some clouds and rain, Mrs. James complained about the heating in her classroom and how she hated to dress in layers. When a teacher told her about a recent trip to Disneyland with her family, Mrs. James made sure to note how overpriced and commercialized the park was, and how dreadful the lines were there. On one particular day, when the principal handed out chocolates to the entire staff to show his gratitude for our hard work, Mrs. James complained about the calories and said that our staff lounges were contributing to obesity.

Mrs. James probably had a limbic system that set an emotional tone of negativity for her. No matter what happened, her limbic system filtered all her interactions and conversations through negativity and sadness.

After you have slept, have felt motivated to get out of bed and have set the emotional tone for the day, you probably then decide whether or not you want some breakfast. The limbic system is also involved here, as it controls your appetite levels. This is why when people are depressed, they may eat too much or too little.

As you leave the house and head for work or school, the deep limbic system is involved in how you bond and connect with people. This is a critical component of our lives, because our joy, feelings, mood and emotions are dependent on our ability to connect with other people on a personal and social level. The simple fact is that humans are social animals. We do not function well when we are isolated and alone. When we develop strong bonds and relationships with others, we feel better.

As you go about your day, chances are things will happen that trigger some particularly charged memories, good or bad. The deep limbic system is also at work here, because it is the part of the brain that stores these highly charged emotional memories. Additionally, the collection of these highly charged emotional memories are what help set the overall tone for your limbic system and mood. The more trauma and negativity we have faced, the more likely we are to store these highly charged memories and then filter new information in a more negative way.

Have you ever smelled something that triggered a flood of emotions and memories of something in your past, changed your mood and even affected your school or work performance? This is due to the fact that the olfactory bulb, which processes smells, is also located within your deep limbic system.

Problems with the Deep Limbic System

Generally speaking, when the deep limbic system is less active, a person may feel more positive and hopeful. Initially, it might seem like the opposite would be true. In fact, I have had many students ask me, "So if my deep limbic system is working hard, doesn't this mean that I should be more passionate and better connected and bonded with other people?" This is not the case. Instead, when our deep limbic system is *overactive,* we have the tendency to become overly emotional, negative, sad, isolated and depressed.

Research has shown that when the deep limbic system in animals is damaged, the parents are not able to bond with their young. One study on rats found that mothers with a damaged deep limbic system dragged their offspring around the cage as if they were inanimate objects.

For humans, an overactive deep limbic system can be problematic in many ways. Typically, an overactive deep limbic system manifests itself as an increased sense of negativity and sadness. Such individuals will interpret harmless or even joyful events as negative and sad. Think of Debbie Downer. Think of Mrs. James. An overactive deep limbic system can also manifest itself as a decrease in self-esteem and self-worth. Individuals feel like they cannot accomplish anything well, and then feelings of hopelessness, helplessness or worthlessness set in.

Changes in sleep patterns and eating can also occur when the deep limbic system is overactive, and some people experience feelings of guilt. There is often a decrease in motivation, drive and libido. People with a struggling deep limbic system often lose interest in activities that were normally exciting and engaging before. They may wake up in the morning and feel like there is no point in getting out of bed. Their energy is low, and social connections suffer. These people often withdraw from others and become more isolated and alone.

Marcie came into my office at the Amen Clinics, where I worked interviewing patients who were being treated for brain struggles. My job was to collect all the information I could about her experiences,

continued on page 64

background, struggles and symptoms so that I could compile a report for her treating psychiatrist. Marcie was in her late fifties, and the exaggerated dark circles under her eyes hinted that she had been through her share of struggles and sadness.

Over the next two hours, Marcie sat in the chair across from my desk and spoke in a polite, but solemn, manner. As we talked, Marcie discussed how depressed she had been feeling. Sadly, this was not new for her. She had been struggling with symptoms of depression her entire life. She had finally decided to seek help because she had reached the point where the idea of being dead seemed more compelling than being alive.

As she gave me the details of her life, I couldn't help but feel her pain and sadness. She was an only child in a family of only children. Her dad was in the military his entire life, so they had moved around quite often. This had made it impossible for Marcie to develop any lifelong relationships or meaningful friendships as a kid.

When she was a young woman, she met a man who was also in the military. He, too, was an only child from a family of only children. And his dad was in the military, and they had moved around quite a bit, too. They quickly fell in love and married. Over the course of their thirty years of marriage, her husband remained in the military, and they were required to move often. They had decided not to have kids, which meant that their home was fairly quiet and peaceful. At the same time, Marcie never felt like she was able to develop any friendships with other people.

When she was in her forties, her husband's parents passed away. In her early fifties, she lost her own parents. Then, two years prior to her appointment with me, her husband passed away from an aggressive form of cancer.

As Marcie recounted these events, she started to weep. After a few moments, she looked up at me and said, "I am literally all alone in this world. My husband is gone. All of his family is gone. My family is all gone. I have absolutely no family alive. And I have no friends.

I could literally die today, and there would not be a single person on this planet who would notice."

At this point, my eyes began to well up with tears as I tried to imagine the isolation, sadness and pain she was feeling. I even thought about taking this lady out to dinner with my family to let her know that there were, indeed, people in this world that she could call friends.

Because she had already had a brain scan performed the previous day, I was able to show her how the deep limbic part of her brain was working way too hard. She was shocked to see it lit up on her scan like a lightbulb. Then, as we detailed the behaviors, symptoms and struggles associated with the deep limbic system, Marcie wept some more.

She realized that her brain was wired in a way that was preventing her from feeling joy. She began to see how her social isolation, low motivation, poor sleep, lack of appetite, and feelings of hopelessness and helplessness were all constructs of brain function.

However, instead of feeling empowered by this information and realizing that she could make a change for the better, her deep limbic system made this realization even more negative. She became even more depressed that she hadn't sought help sooner.

Conditions Associated with the Deep Limbic System

In general, symptoms of an overactive deep limbic system include feelings of sadness or unhappiness. Overactivity of the deep limbic system is often associated with depression, mood disorders, pain syndromes and dysthymia (chronic mild depression). Sometimes, symptoms are so severe that a person knows something is wrong. Other times, however, a person might feel miserable but not be able to articulate why.

Other symptoms of an overactive limbic system include irritability and frustration, even with small things. There is a loss of interest or pleasure in normal activities. There are changes in sleep and appetite, feelings of agitation or restlessness, slowed thinking, indecisiveness, difficulty with concentration, crying spells, and feelings of hopelessness, helplessness and worthlessness.

For teenagers and young adults, symptoms might also include anxiety, anger and avoidance of social interactions. Oftentimes, depression occurs with other behavioral problems and brain struggles, such as anxiety disorders or ADHD. Performance in school or work might also suffer.

Some of the most dangerous symptoms associated with an overactive deep limbic system are suicidal thoughts and ideations. When a person starts to feel depressed to the point of thinking about death and/or suicide, this should *never* be taken lightly. Help should be sought immediately.

What Does It All Mean?

There is a significant difference in the deep limbic system of males and females. Research has shown that females have a larger deep limbic system than males.

Because the deep limbic system is larger in women, bonding comes easier to them. Generally speaking, women have a greater nesting instinct, and they are typically the primary caretaker in the home. This is not to say that women have to be moms or that men are not capable of being the primary caretaker. It just means that when we look at careers that focus on the care of children and the elderly, there is a reason why these jobs are easier for women— they have a larger deep limbic system.

Another positive for women of having a larger deep limbic system is that they are typically more likely to seek help when there is a problem, and they are much more likely to think of options for healing.

However, having a larger deep limbic system can also have its drawbacks. Studies and statistics have repeatedly shown that women experience a greater incidence of sadness and depression. They are twice as likely to attempt suicide than men. Women who do so, however, have often reported that their ultimate goal is not to end their life, but to get themselves and others to realize it is time to get help.

Another unfortunate reality is that even though women are twice as likely to attempt suicide than men, men are actually three times as likely to be successful at it. When men feel depressed, they are less likely to seek help and look at the options that are available to them. Instead, they decide quickly to end their life, and then follow through with this course of action.

Too often, I have sat with students who have struggled with depression and who blame themselves for this. They feel sad, unmotivated and uninterested in much of what life has to offer. They hate that they feel this way, but then they feel hopeless and helpless to do anything to change it. Ultimately, they hate themselves because they are not making the effort to make things better, even though their lack of energy and drive to make things better is diminished by the very same disorder. It is a vicious cycle of self-loathing and hopelessness that keeps people feeling low and worthless.

The good news is that once you understand the dynamics of this interplay of emotions, mood and motivation within the deep limbic system, you can see the true power you have to change your brain and change your life. Part III will detail some of the exciting and powerful ways people can optimize the deep limbic system to minimize sadness and break the cycle of depression.

SUMMARY	
Deep Limbic System Functions	**Deep Limbic System Problems**
» Mood control	» Depression, sadness
» Charged memories	» Focus on the negative, irritability
» Modulates motivation	» Low motivation and energy
» Sets emotional tone	» Negativity, blame, guilt
» Appetite/sleep cycles	» Poor sleep and appetite
» Bonding	» Social disconnections and isolation
» Sense of smell	» Low self-esteem
» Fight-or-flight response	» Hopelessness
	» Decreased interest in things that are usually fun
	» Feelings of worthlessness or helplessness
	» Feeling dissatisfied or bored
	» Crying spells

Having high DLS activity may entail some positive traits, such as:	
» More in touch with feelings	» Increased empathy for people who are suffering

Some medical conditions associated with high DLS activity are:	
» Depression	» Dysthymia (chronic mild depression)
» Cyclic mood disorders	» Pain syndromes

Common treatments implemented when DLS activity is high:	
» Learning to fight our Automatic Negative Thoughts (ANTs)	» Intense aerobic exercise
» Biofeedback	» Relationship counseling
» Cognitive-behavioral strategies	

QUESTIONS

1. How do you think your deep limbic system is operating?

2. What can you relate to in this chapter? Does the information in this chapter shed light on the behavior of someone in your life?

3. What are some things you might do differently in your own life now that you know about the deep limbic system and what it controls?

6 ANXIETY AND THE BASAL GANGLIA

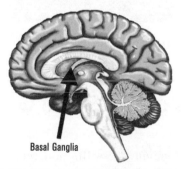

FIGURE 6.1: **THE BASAL GANGLIA**

The basal ganglia are the last major part of the brain that we will focus on. They are located deep in the brain, surrounding the deep limbic system. In fact, the basal ganglia and the deep limbic system are interconnected, which means that they often work together to ease stress, or against each other causing serious issues.

The images below show a comparison between two people's brains. The scan on the left is of a brain that is healthy. You will notice that there is overactivity only in the bottom of the brain, which is normal for the cerebellum. On the right scan, however, you can see that the basal ganglia are working too hard. This person was struggling with post-traumatic stress disorder (PTSD), which began after military service in the war in Afghanistan. I'll talk more about this condition later in this chapter.

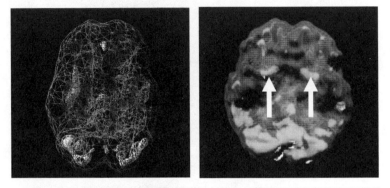

FIGURE 6.2: **HEALTHY BRAIN**   FIGURE 6.3: **OVERACTIVE BASAL GANGLIA**

Functions of the Basal Ganglia

From a physical perspective, the basal ganglia are largely involved with movement, including coordination, muscle tone, tremors, walking and voluntary movement. They have been associated with Parkinson's disease, which is characterized by shaking, muscle rigidity and difficulty standing or walking; with Huntington's disease, which is characterized by odd jerking movements and a loss of memory; and with cerebral palsy, which is characterized by various motor problems and low muscle tone.

More recently, abnormal activity in the basal ganglia has been linked to behavioral problems and mental disorders as well. People with normal activity in the basal ganglia typically have a nice balance of anxiety and stress. Believe it or not, anxiety is a necessary part of life. When you are in school, anxiety is what helps you do your best on exams, reports and presentations. It keeps you from procrastinating too much and serves as a reminder of the consequences of not being focused or on task.

Some anxiety is also important in our jobs. It helps us work hard to please our supervisors and customers, and it keeps us from calling in sick or playing hooky because we stayed out too late the night before.

We can also thank the basal ganglia for providing us with healthy levels of stress whenever we make life-changing decisions. Thinking about buying a new car, making an investment, selecting a university, buying a house or proposing to that significant other? The basal ganglia are what step in with angst and make you think long and hard about the long-term impact of your decision.

A healthy amount of anxiety is also necessary to keep us from physical danger. If we see a house on fire and realize that we left our laptop inside, the basal ganglia tell us, "Hmmm, probably not a good idea to run into that burning building to retrieve the laptop. I don't really feel like burning to death today."

Problems with the Basal Ganglia

When the basal ganglia work too hard, a person might have a tendency to be overly anxious. Instead of having a healthy amount of anxiety about writing a final paper or accomplishing a task at work, this person will stress out over it. This stress can cause physical symptoms, including headaches and stomachaches.

Another hallmark of basal ganglia that work too hard is the tendency to predict the worst in every situation. Individuals with overactive basal ganglia are high stress and high worry, and they tend to imagine all the worst-case scenarios in any given situation. Their argument is that they "want to be prepared for the worst." Some might even feel a sense of pride about this. In reality, however, this behavior occurs because of a brain that is causing unnecessary stress and thus is working against the individual's best interest.

I am a perfect example of a person with overactive basal ganglia. Recently, I went on a trip to Indonesia to conduct some work with professors, teachers and students in a variety of settings. Some of the schools I visited were in areas where disease, poverty and turmoil were rampant.

From the beginning, my basal ganglia were on fire. When the flight itinerary was first emailed to me, I made sure to check out the safety ratings of the airlines and the actual aircraft in which I would travel. I had my doctor write me a prescription for Xanax to ensure that I could get through the sixteen-hour flight. My suitcase was filled with more hand sanitizer, medications, supplements and digestive supports than clothes. In short, I was prepared for the worst.

I traveled with a group of fifteen educators, administrators and board members from my university. As much as I tried to conceal my anxiety, my colleagues quickly picked up on the fact that I was limiting my food intake to steamed rice and bottled water. And no matter how hard they tried to get me to be adventurous and try something new, my mind played out the multiple ways that I would get sick and become miserable. As a result, I became the target of ridicule and jokes from my peers. This didn't bother me, because I would rather face a little ribbing than battle the cold sweats, flu-like symptoms and explosive diarrhea associated with eating something that did not agree with me.

One of the dangers of having overactive basal ganglia that cause unnecessary stress and anxiety is that sometimes our concerns are validated. When something does happen that confirms our worries, we are then fueled by it, which can lead to even more anxiety and irrational fear if they aren't put in check.

This was the case during my trip to Indonesia. The night before the graduation ceremony was to take place at our partnering university, our group gathered for a dinner at a popular restaurant in Jakarta. As had been the case all week, my colleagues eagerly tried out the food, basking in new, exotic flavors, as I sat respectfully by, eating my steamed rice and refusing to drink any water unless it was from a bottle that I had unsealed myself.

The next morning, I was in the front row at the two-hour graduation ceremony with several of my colleagues, while our university president, wearing his full academic regalia, was seated onstage, front and center. During the proceedings, I noticed the president shifting in his seat uncomfortably. About halfway through, I saw beads of sweat gathering on his reddening face, and I quickly realized that he was having digestion issues.

Over the next several minutes, I noticed a few others in our group getting up and leaving for about fifteen minutes at a time. Finally, the president also stood up, excused himself and left the stage for twenty minutes.

That afternoon the joking stopped, and I was suddenly everybody's best friend. My luggage became a dispensary, and my colleagues began to stock up on their own supplies of bottled water and sanitizers.

For me, this experience was both good and bad. Surely I, with my passive-aggressive mind, felt victorious in knowing that my irrational anxiety had been validated. The downside was making sure that I did not let this sense of validation cause my anxiety to escalate even further, to the point that it would overrun my thoughts and emotions, and impede my joy.

Conditions Associated with the Basal Ganglia

As you might have guessed, anxiety disorders are very common. In the United States alone, more than forty million people ages eighteen and older are affected by anxiety disorders. This translates into a little more than 18 percent of the population struggling with an anxiety disorder at any given time.

Panic disorder is one such condition and affects more than six million adults in the United States. The onset of panic disorder is typically in early adulthood and includes sudden and repeated attacks of fear that last for several minutes. During one of these panic attacks, people typically have frightening physical symptoms. Some feel as though they are having a heart attack because of the severity of chest pains, the loss of breath and the overwhelming fear they experience. Panic attacks may involve a fear of disaster, of losing control and even of having another panic attack.

Post-traumatic stress disorder affects more than 7.7 million adults in the United States and can develop at any age. PTSD is typically triggered by an extremely terrifying experience that may have involved physical harm to oneself or to another. With PTSD, people often experience flashbacks, disturbing and fearful thoughts, and bad dreams. They might become easily startled, tense, and overly sensitive and/or angry. Children and teens may regress to bed-wetting, may stop talking, may act out the traumatizing event in play, and may become overly needy or clingy with a parent or other adult.

Generalized anxiety disorder affects nearly seven million Americans and can occur throughout a person's life. Symptoms include constant worrying or obsessing, sweating, nausea, shortness of breath, diarrhea, rapid heart rate, muscle tension, trembling, difficulty sleeping, irritability, fatigue and restlessness. Younger people might notice excessive anxiety related to school, work or sporting events. They may also be overly anxious about being on time, fitting in with peers, being a perfectionist, needing approval and even dealing with natural disasters.

Social anxiety disorders include a heightened level of anxiety about being with other people, which affects a person's ability to interact with others, even when he or she wishes to do so. Those with social anxiety disorders are overly concerned about what people think about them and are constantly worried about being judged. They may spend days or weeks fretting about an upcoming social event. They are self-conscious and will blush, sweat or tremble around others. They will often feel nauseous or sick in social situations, as well.

Specific phobias, which are characterized by a marked and persistent fear and avoidance of a specific object or situation, affect more than nineteen million Americans over the age of eighteen. They often begin in childhood, around the age of seven. Common types of phobias include fear of spiders, dogs, snakes, insects, mice, flying, riding in cars, driving, bridges, tunnels, elevators, storms, heights, water, needles, blood, loud sounds, clowns, monsters, falling down and much more.

What Does It All Mean?

Over the last four chapters, you have been given quite a bit of information about four specific parts of the brain, the behaviors they are involved with and what happens when things go wrong. No doubt, you have probably found yourself connecting with many of the behaviors and symptoms delineated in these pages, and you may have even started diagnosing which parts of your brain might be underactive or overactive.

Keep in mind that we all relate to many of these behaviors and symptoms at some point or another. When we think about the prefrontal cortex, for instance, all of us can come up with plenty of times when we were impulsive, when we procrastinated, or when we had difficulty with concentration, focus, attention and organization. When we consider the cingulate, we can always recall situations where we had trouble letting go of thoughts or behaviors or where we felt stubborn and oppositional. And when we contemplate the deep limbic system, we can all identify a time in our lives when we felt particularly down or depressed, or lacked motivation and drive.

The basal ganglia are no different. It is perfectly natural and commonplace to be anxious and fearful at times, whether it is with friends or significant others, in school, at a job or in other social situations.

Though you may identify with many of the struggles you have read about up to this point, it is important to know what separates normal from needing help. This is especially true for those of you who may have overactive cingulates and now find yourselves obsessing about the idea that something is wrong with you. And it is true for those of you with overactive basal ganglia or an overactive deep limbic system who are now convinced that you are going to spiral out of control and fall into a pit of despair. Of course, if you are a reader who is struggling with a prefrontal cortex, you may have forgotten much of what you have already read, because you were distracted by something else.

To a certain degree, it is normal to experience many of the symptoms and struggles that are outlined in Part I of this book, but if any of these symptoms or struggles become long term, or if they are preventing you from feeling joy, succeeding or accomplishing your goals, then it might be time to seek some outside help. No matter what, it is important to remember that your struggles are common and are a result of blood flow and brain activity. They have nothing to do with you being a bad person or doing something wrong. Each of the behaviors, symptoms, emotions and feelings that we have covered in Part I have one thing in common: brain function.

We often attribute behaviors and struggles with the people them-selves and may even criticize them for it. We judge people who lack empathy and are impulsive, and we label them jerks. We tell those who are struggling with obsessions and compulsions to "just stop!" We assume that individuals with depression are depressed because they want to stay that way, and that if they just started doing things again, they would naturally get better. We even joke about people with anxiety and assume that they can easily put their irrational fears and angst to rest if they just try hard enough.

Now that you have built a foundation of understanding of the human brain, the hope is that you can see human behavior, mental disorders and brain struggles in a new light. Telling a person with an underactive prefrontal cortex to just try harder, a person with OCD to stop obsessing, an anxious person to stop worrying or a depressed person to be happy is pointless.

However, it is essential to keep in mind that if you are struggling with certain regions of your brain, it is futile to use your struggles as a crutch. You might have ADHD, depression, anxiety or OCD, but this doesn't mean that the world needs to adapt to your needs. Knowing what you know about the brain now means that it is up to you to take control of your brain and your life.

One of the most exciting aspects of learning about the brain is the realization that we have the power to make a change. Knowing what we know empowers us to improve our brains so that we can optimize our potential and minimize our struggles.

In Part II of this book, you will see how every decision you make every day of your life has the potential to help your brain or harm it.

SUMMARY

Basal Ganglia Functions	Basal Ganglia Problems
» Integrates feelings, thoughts, movements	» Sets anxiety levels
» Sets body's idle, like a car engine's	» Hypervigilance
» Smooths movement	» Muscle tension
» Modulates motivation	» Conflict avoidance
» Mediates pleasure	» Predicts the worst
	» Excessive fear of being judged by others
	» Tendency to freeze in anxious situations
	» Seems shy or timid
	» Bites fingernails or picks skin
	» Excessive motivation, can't stop working
	» Panic
	» Physical stress: headaches, stomachaches

Having high BG activity may entail some positive traits, such as:	
» Increased motivation	» Ability/desire to work for long periods
» Conscientious	» Self-discipline

Some medical conditions associated with high BG activity are:

» Anxiety disorders

» Physical stress disorders, such as headaches or gastrointestinal problems

» Insecurity

Common treatments implemented when BG activity is high:

» Learning to kill the ANTs	» Intense aerobic exercise
» Body biofeedback, hypnosis, meditation	» Relaxing music, assertiveness training
» Limit caffeine/alcohol	

QUESTIONS

1. How do you think your basal ganglia are operating?

2. What can you relate to in this chapter? Does the information in this chapter shed light on the behavior of someone in your life?

3. What are some things you might do differently in your own life now that you know about the basal ganglia and what they control?

Part II

The Young Brain in the Real World

# THE DEVASTATING IMPACT OF DRUGS AND ALCOHOL

I hope you have now developed a deeper understanding of and appreciation for the human brain and how many of our behaviors and struggles in life can be attributed to the way certain parts of our brain are functioning at a given time.

Now we're going to talk about the specifics of how the young brain develops and interacts in the real world, with a particular focus on the countless ways that people can do damage to their brain, intentionally or not.

Illegal Drugs

I often ask students about the many ways that people can damage their brain. During this exercise, as they list all the ways they can think of, students are usually engaged and excited. Many times, their first answer is drugs. They aren't wrong. Using drugs, such as cocaine, heroin, methamphetamines and marijuana, is indeed an obvious way that many people damage their brain.

Take a look at the brain scans in the figures below to see the true impact of drugs on the human brain.

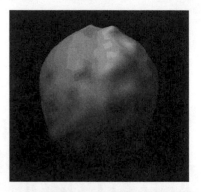

FIGURE 7.1: **HEALTHY BRAIN**

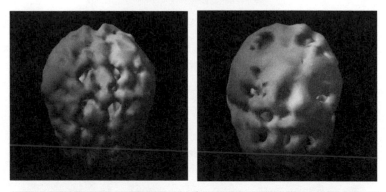

FIGURE 7.2: **COCAINE**   FIGURE 7.3: **METH**

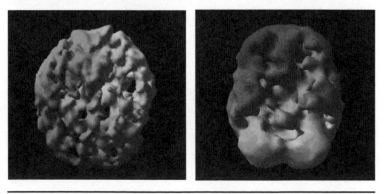

FIGURE 7.4: **HEROIN**   FIGURE 7.5: **MARIJUANA**

The images you see here are actual surface-view brain scans of individuals. In the scan of a healthy individual, you can see how full, even and symmetrical the brain looks. In the other scans, you can see "holes" in the brain—areas of underactivity and dysfunction. I think we can all agree that these drugs are devastating to the brain.

The effects of drug use are even more significant in the young brain. Before the age of twenty-five, your brain is still developing at a phenomenal rate, and how your brain develops during this time is likely to determine the success you achieve during the rest of your life. Any time you introduce toxic substances into your body, their effects can disrupt brain development and cause damage that will last a lifetime.

More and more research is confirming the fact that drug use, while terrible at all ages, is more disastrous the younger you are. One study found that young drug abusers were up to three times more likely to suffer brain damage than those who didn't do drugs. In studies of teenagers and young adults who used heroin, damage to the nerve cells involved in learning, memory and emotional well-being was found to be similar to that in people in the early stages of Alzheimer's disease.

Club drugs, such as Ecstasy and methamphetamines, have been found to inflict damage on the brain similar to that caused by a traumatic brain injury. These drugs trigger chemical chain reactions in the brain that cause cell death, memory loss and irreversible brain damage. Studies of cocaine users have shown that their brain function (even when they are not actively using cocaine) is sluggish and abnormal when compared to the brain function of nonusers. The damage inflicted by cocaine use leads to memory loss, learning problems, attention deficits and even strokes.

This doesn't mean that starting to use drugs after the age of twenty-five is without consequence. Using drugs later in life will also result in some degree of brain damage.

Interestingly, there is a reason that people typically do not start using drugs after the age of twenty-five, and that is that by then they have a fully developed prefrontal cortex. So instead of acting on impulse or feeling the deep limbic need to fit in with peers, the individual with the mature brain is far more able to assess the situation and decide that it is probably not in his or her best interest to use drugs that could devastate the brain and the potential for success in life.

The irony here is that young people are at the greatest risk of damaging their brain by using drugs and alcohol, yet the very part of the brain that is most likely to keep them from engaging in such behavior is the part that is still not fully developed.

Marijuana

These days, marijuana (cannabis) seems to be a hot-button topic. Over the years, mainstream American culture has accepted the use of marijuana to a greater and greater degree, and states have started to legalize marijuana for both medicinal and recreational purposes. The first major milestone in the fight for marijuana acceptance took place in 1996, when California legalized marijuana for medical use. Since then, the momentum toward total legalization has only increased.

When I show the brain scans of people who use drugs in my classes, the marijuana slide is the one that gets the most passionate responses. Several years ago, the argument was, if marijuana is bad for you, then why are people using it for medicinal reasons? My answer back then was that it was being used to ameliorate the pain associated with AIDS, cancer, and other serious, painful and terminal diseases. In such cases, how relevant was the damage done by marijuana when the individual was already headed toward death? If marijuana can make seriously ill people feel better and live more comfortably, I am all for it.

The real debate about marijuana here is whether healthy young people—who are still in the process of crucial brain development—should use marijuana. Rather than get pulled into the political debate about marijuana, let's focus on what we know about its effects on the human brain. First and foremost, let's answer the most critical question: Does marijuana damage the brain? **Yes.**

Over the years, research has illustrated repeatedly that marijuana damages the brain, particularly the memory and learning centers of the brain. More recent research has also found that the earlier people start using marijuana, the worse the brain damage is.

One 2012 study conducted in Melbourne, Australia, was able to identify the impact marijuana has on the white matter of the developing brain. Think of white matter as the network cables that transmit information across the computers (the gray matter) within your brain. In the case of marijuana users, researchers found disruptions in the white matter fibers. This means that marijuana often leads to memory impairment, impairments in learning and difficulty in concentration. Even more frightening was the discovery that there was more than an 80 percent reduction in white matter in the brains of marijuana users, and the damage increased significantly the younger the person was when he or she started using this substance.

That said, every study about the dangers of marijuana seems to be countered by other studies showing that it is harmless. In fact, some studies have even been published that claim marijuana can actually be helpful for the brain. One such 2012 study out of Tel Aviv concluded that tetrahydrocannabinol (THC), which is the main psychoactive component of marijuana, enhanced biochemical processes in the brain, meaning that the brain cells were more protected and were preserved over time. The study also found that marijuana increased performance on learning and memory tests. As you can imagine, after this study was released, the media was flooded with news articles about the potential health benefits of marijuana.

Several media outlets published reports of a 2007 study out of Switzerland that seemed to show that marijuana was actually beneficial to teens. These reports claimed that the grades of teenage marijuana users were comparable to those of their peers who did not use marijuana, they were just as likely to finish school, they had more friends and they were more likely to participate in sports than teenagers who did not use marijuana.

Unfortunately, there is a reason why recent polls have highlighted the fact that Americans' distrust of the media continues to hit new highs. A recent 2012 study found that 60 percent of Americans have little to no confidence in the mass media's ability to report the news fully, accurately and fairly. Sadly, this is due to the fact that many news outlets focus on the headlines instead of the underlying facts and details.

In the case of the Tel Aviv study, the media was accurate in reporting a protective benefit of THC. What the media failed to report, however, was that the amount of THC injected into mice in the study was limited to a single very low dose—up to ten thousand times *less* than the amount of THC found in a typical joint. Additionally, the study showed that this trace amount of THC actually resulted in minor damage to the brain at first, which, researchers then theorized, may have preconditioned the brain to protect itself against more severe damage in the future. Does this mean that teenagers or young adults who use marijuana are actually helping themselves by protecting their brain from further damage? Absolutely not.

As for the Swiss study, while much attention was paid to the fact that it found few differences in the grades and dropout rate of marijuana users compared to their peers who were nonusers, the study also showed that youths who used both marijuana and tobacco did the worst in school, had the poorest grades, were less likely to finish high school, were more likely to be depressed and were more likely to get drunk frequently. And those who used marijuana but did not smoke tobacco were found to skip school more and to have more sensation-seeking personalities, making them more likely to use alcohol and other drugs, than teenagers who did not use marijuana at all. It is important to note that the study focused only on correlations and associations, so it was impossible for the researchers to show that using or abstaining from marijuana was the specific cause for anything they reported.

The bottom line here is this: Before you make the decision to start (or to continue) using any type of substance, consider its potential deleterious effects. It should be clear by now that your brain is the most critical organ of your entire body. When your brain is working right, you have the best chance of achieving success and being the best you can be. When your brain is not working right, you end up struggling and having difficulties in your life.

Why would you want to disrupt your brain by introducing toxic substances? It would be like pouring salt into the gas tank of a Ferrari!

The brain scans we have taken of marijuana users confirm this. Following are four illustrations of the damage that marijuana inflicts on the human brain:

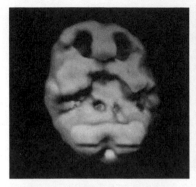

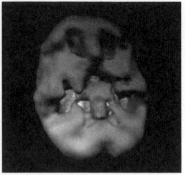

FIGURE 7.6: **BRAIN SCAN OF AN EIGHTEEN-YEAR-OLD WITH A THREE-YEAR HISTORY OF MARIJUANA USE FOUR TIMES A WEEK,** underside surface view, decreased prefrontal cortex and temporal lobe activity

FIGURE 7.7: **BRAIN SCAN OF A SIXTEEN-YEAR-OLD WITH A TWO-YEAR HISTORY OF DAILY MARIJUANA USE,** underside surface view, decreased prefrontal cortex and temporal lobe activity

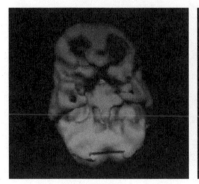

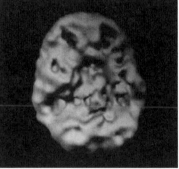

FIGURE 7.8: **BRAIN SCAN OF A THIRTY-EIGHT-YEAR-OLD WITH A TWELVE-YEAR HISTORY OF DAILY MARIJUANA USE,** underside surface view, decreased prefrontal cortex and temporal lobe activity

FIGURE 7.9: **BRAIN SCAN OF A TWENTY-EIGHT-YEAR-OLD WITH A TEN-YEAR HISTORY OF MOSTLY WEEKEND MARIJUANA USE,** underside surface view, decreased prefrontal cortex and temporal lobe activity

As you examine the different brain scans of individuals who have used marijuana, cocaine, meth and heroin, ask yourself the following: Do you think these people are able to achieve their full potential? When you see how their brains have been affected by their drug use, do you believe they are truly themselves? More important, which brain do you want?

Alcohol

The consumption of alcohol by teenagers and young adults is also a fairly complicated and sensitive issue, especially since the use of alcohol (unlike marijuana) is legal at age twenty-one. But like the other substances I've talked about, the harsh reality of alcohol is that it is damaging to the brain—the young brain, most of all.

One particular 2012 study out of the University of California, San Diego, and the University of Pittsburgh looked at the effects of alcohol on the brain of sixteen- to twenty-year-olds over a period of eighteen months. The researchers discovered that those who had consumed five or more alcoholic beverages a couple of times a week lost a significant amount of white brain matter, which is associated with attention, decision making, judgment, self-control and memory. Even worse, these impairments were likely to last into adulthood.

When you contemplate what it is like to be drunk, it is obvious that alcohol damages the brain. When a person is drunk, he or she has slowed reaction times, slurred speech, difficulty walking, blurred vision, impaired memory and poor judgment. These are all connected to brain function. Alcohol is bad for the brain because it lowers overall blood flow to your brain. Your brain needs blood flow to function well. Over time, a continual decrease in blood flow can diminish a person's memory and judgment permanently.

The brain scans you see in Figure 7.10 below are from the same person. The scan on the left was taken when this individual showed up at the Amen Clinics for his first appointment. The scan on the right was taken when he came back the next day, drunk. Notice the overall decrease in blood flow and activity in the brain—just from alcohol.

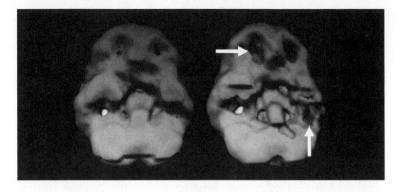

FIGURE 7.10: **SOBER BRAIN AND DRUNK BRAIN**

In addition, drinking alcohol daily has been shown to increase the risk of Alzheimer's disease. Newer research has also illustrated that even the moderate consumption of alcohol—three to four drinks a week—can have negative effects on the brain. This might be because alcohol works to dehydrate the brain. In fact, a 1995 study found that people who drank alcohol just a couple times each week had a physically smaller brain than nondrinkers. I don't know about you, but I don't think having a smaller brain is good for you.

While researchers have suggested that having a glass of wine each day is good for your heart, you have to stop and ponder which organ is more important. Your heart? Or your brain, which controls your heart and everything else?

Reasons for Using Alcohol and Drugs

We have all been through the classes about substance and alcohol use and abuse at school, heard the horror stories and listened to guest speakers discuss how bad alcohol and drugs are for young people. We have made pledges to resist drugs, signed ribbons to abstain from alcohol, seen commercials of a brain on drugs being compared to eggs in a frying pan and watched videos that teach us how to stand up to peer pressure.

When you read through the research on drug and alcohol use and see the actual brain scans of individuals who have been using and/or abusing alcohol and drugs, it is clear how toxic these substances are for the human brain. I have talked with teenagers and young adults extensively on this subject, and I am certain that they do understand that these substances are inherently bad for the brain, body and life.

So, if young people have been inundated with information about the destructive effects of drugs and alcohol, and if they know deep down that drugs and alcohol are bad for them, the real question becomes, Why do young people still use alcohol and drugs?

As you might expect, the answer to this question is complicated. A number of factors must be taken into consideration. One such factor is dopamine, a chemical neurotransmitter that is released by nerve cells to send signals to other nerve cells. Dopamine helps control the brain's reward and pleasure centers by responding to these rewards and by motivating us to take action to receive them again. Think of it this way: when dopamine is released into our brain, we feel pleasure and a sense of being rewarded. When this occurs, we seek that pleasure and reward even more. And, as I discussed in Chapter 2, our reward and pleasure centers hit their peak of development right in the middle of our teenage and young adult years.

Drug and alcohol use floods our limbic brain with dopamine. Drugs have the potential to raise the amount of dopamine in our brain up to ten times its normal level. When this occurs, our brain begins to perceive that the elevated levels are normal, and then addiction develops as a way to maintain those higher dopamine levels.

Sex, video games, technology, food and exercise have also been found to release dopamine into your brain and make you feel pleasure and a sense of reward. And, just as with drugs and alcohol, as the brain becomes more accustomed to higher levels of dopamine, the risk of addiction to these activities increases. This is why people may feel that socializing with friends is boring when no alcohol is involved. This is why people can also feel bored when they are not texting, surfing the Internet or playing video games. It's all about the brain.

Teenagers and young adults are more susceptible to using drugs and alcohol because their brains are still developing. The prefrontal cortex should step in and say, "Although consuming this alcoholic beverage or using this drug might make me feel better in the short term, perhaps it's not the best idea for me, because of its toxic effects on my brain and the potential for damage." Yet, as you already know, the prefrontal cortex is typically one of the last areas of the brain to develop fully. This means that the most important part of our brain isn't fully on duty exactly when it is needed most to stop you from doing damage to your still developing brain.

Because our deep limbic system is also in active development during these years, we are more driven by emotion than logic. We are also driven by peer pressure to engage in activities that we think will bring us social praise and/or acceptance. But don't forget that, as mentioned in Chapter 2, peer pressure comes more from our own desire to please others than it does from other people.

Self-Medication

While there is a host of reasons why teenagers and young adults use alcohol and drugs, among them social acceptance and peer pressure, there is one major cause that is often overlooked and misunderstood: they like the way alcohol and drugs make them feel. Over the years, I have talked with countless teenagers and young adults, and this is the most common reason they give for turning to drugs and alcohol.

Alexis is the perfect example of this. Alexis was twenty years old when she came into my office for evaluation. She had been struggling with anxiety and depression for as long as she could remember, and she had developed an addiction to alcohol and marijuana as a result.

When Alexis was in elementary and middle school, she felt anxious, nervous and sad almost all the time. She was often overly concerned about what her peers thought about her, and she would freeze in social situations. She had a hard time completing her work at school because she was too distracted and had difficulty concentrating and focusing.

In the eighth grade, she tried marijuana for the first time. She described this experience as getting a glimpse of what heaven must be like. For the first time in her life, she said, she experienced what it must feel like to be normal. Her anxiety subsided, she could focus and she suddenly became more sociable and friendly.

As you can imagine, her consumption of marijuana increased dramatically after that. Alexis became a better student and had more friends all throughout high school. Along the way, she began drinking, as well. Just as was the case with marijuana, she found that she really enjoyed alcohol because it made her feel relaxed, carefree, social and happy.

Of course, Alexis did not stop to consider the long-term effects of her dependence on alcohol and marijuana. In fact, she assumed that nothing was being damaged, because her grades actually improved. During her second year in college, however, Alexis noticed a definite shift in her use of alcohol and marijuana. She realized she was under the influence far more than ever before, and her academic performance began to decline significantly. She found herself struggling with her memory and her ability to keep organized and focused.

In the figures below, you will see two scans that show the inside workings of the brain. In these scans, the white areas of the brain indicate the hardest-working parts of the brain. The scan on the left is of a healthy brain. The only areas of white are toward the bottom, in the cerebellum, which is normal since over half of the brain's neurons are located in the cerebellum. The scan on the right is Alexis's brain in its natural state, without drugs or alcohol.

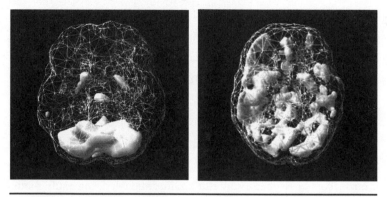

FIGURE 7.11: **HEALTHY BRAIN**   FIGURE 7.12: **ALEXIS'S BRAIN**

As you compare these two brain scans, it suddenly becomes pretty clear why Alexis enjoyed using alcohol and marijuana—her brain was on fire! Look at the overactivity in all the different regions of her brain. She was having difficulty with anxiety, focus, sadness and socializing with her peers because her brain was clearly working against her.

It is important to understand why alcohol and marijuana were her drugs of choice. What does alcohol do to the brain? It slows it down and acts as a depressant. What does marijuana do to the brain? It also slows and quiets the brain. Alexis said she liked using alcohol and marijuana because it made her feel, work and function better. Essentially, she was self-medicating a brain that was clearly struggling. When she was using, her brain was working better.

Unfortunately, Alexis's choice of self-medication worked against her: over the course of many years it did damage to her brain. And while she felt as though she was enhancing her performance and sense of well-being through the use of alcohol and marijuana, she was actually introducing toxic substances into her body at the expense of her long-term potential for success and normal brain function.

As I sat down with Alexis and reviewed her brain scans, she was able to make this connection for herself. "Well, lovely. No wonder alcohol and marijuana make me feel good. My brain looks like a freaking light show up in there." We talked about better techniques to make her feel good, ones that would not damage her brain. You will read about this in Part III of this book.

What Does It All Mean?

Now that you have learned about the brain's amazing complexities and can see how your brain is responsible for who you are, how you act, how you interact and your ultimate potential for success and joy in life, my hope is that you have a newfound appreciation for the human brain.

At the same time, I hope this knowledge about the brain's fragility also terrifies you enough to take special care of this organ. Knowing what you know about the brain now, why would you want to damage the most important part of what makes you who you are?

It comes down to this: If you find that you need to use alcohol, drugs or any other type of substance to make you feel good, then you should be concerned. It probably means that your brain is struggling.

Why is it that some people are happy when they are drunk, but others are angry and violent? Why do some people get sleepy when they consume alcohol, yet others become energized and sociable? There is a reason why some people hate the way drinking alcohol makes them feel.

Likewise, some people who use marijuana might hallucinate or become paranoid. Others will use marijuana and feel more relaxed and sociable. Still, others might use marijuana because they feel it actually enables them to function better. And some might use marijuana with very little effect.

The reason for all these differences is that we all have a brain that is unique to us. Each of our brains works differently than another person's brain. As a result, we each react differently to whatever we put into our bodies. One person can use Ecstasy for years and feel like there haven't been any permanent effects. Another person can use Ecstasy one time and die.

If there is only one thing you remember from this chapter (or from this book, for that matter) it is this: **self-medication is dangerous and unnecessary.** Also, know that alcohol and drug consumption, especially if done while your brain is still developing, is *always harmful.* There is no escaping this, even if you don't feel the negative effects of alcohol and drug consumption right away.

If you have a brain that is underactive or overactive, and you find yourself struggling in various aspects of your life, there is plenty that you can do naturally to help your brain work better. If you find yourself truly struggling, there are plenty of options, including supplements and medications, that can make you feel better—without all the damage to your brain. We'll talk more about what you can do to optimize your brain in Part III.

SUMMARY

- Drugs, whether legal or illegal, can have a profound impact on your brain function.
- Illegal drugs damage the brain.
- Marijuana has repeatedly been shown to hurt the brain.
- People often use drugs/alcohol as a form of self-medication.

- You can feel great, have fun and have friends without resorting to drugs or alcohol.

QUESTIONS

1. Why do you think some people assert that the use of alcohol, marijuana and other drugs is acceptable and/or harmless behavior?

2. Do you know somebody with an alcohol/drug problem? If so, how has this affected the person and the people around him/her?

3. What are some effective ways to talk with someone who is battling an alcohol or drug problem?

8 OTHER WAYS WE HURT THE BRAIN

When we think of all the different ways that people hurt their brains, it is easy to focus on things like using alcohol, tobacco and drugs. Each of these activities has an undeniable and profound impact on how the brain works. Unfortunately, the ability to harm our brain extends much farther beyond the obvious.

Teenagers and young adults in particular have an amazing amount of power to do harm or good to their brain each and every day. All the choices they make, and even the thoughts they have, matter.

Physical Trauma

As you learned in Chapter 1, concussions and physical trauma to the brain can have a debilitating effect on a person's life. These are typically classified as traumatic brain injuries (TBIs). The Centers for Disease Control (CDC) estimates that nearly two million Americans suffer a TBI each year. Common causes include falls, car accidents, sports injuries and assaults.

Interestingly enough, text messaging while walking or driving has increasingly been cited as the root cause of traumatic brain injuries, especially in teenagers and young adults. In fact, the impairment associated with using a cell phone while driving—whether texting or talking—has been found to be as significant as when a person drives while drunk! Even talking on the phone through Bluetooth technology has been shown to significantly reduce reaction time behind the wheel, resulting in a twofold increase in the risk of a rear-end car accident.

What this means for the developing brain is that you must be more cognizant of the potential dangers that may arise from the choices you make. Sure, playing soccer and football might be engaging, might contribute to your physical health and might even score you a college scholarship. But in the end, is it worth the possibility of sustaining brain damage? These are important questions to consider carefully if you are a teenager or a young adult, because the very act of considering them forces you to activate your prefrontal cortex to look at the future—at consequences—and then make judgments and decisions that are in your best interest.

Consider for a moment the financial success enjoyed by pop icons, athletes and movie stars. Many of them, especially athletes, make a lot of money while they are young. However, many of those who find wealth early on end up struggling, broke and unhappy once they've exhausted their physical body. Then they struggle to make good decisions and feel joy because of how much they beat up their brain when they were younger. The evidence suggests that as many as one third of professional football players will experience some cognitive impairment in their lifetime, including post-concussion syndrome, which can cause severe depression. People often idolize athletes, artists and actors who are wealthy and are living only in the present, but what happens later in life, assuming they even survive beyond their initial fame and fortune?

When I worked in a nondirective play therapy program at a public school district in San Diego County, we were fortunate enough to have several of the pro football players from the San Diego Chargers visit the school. The purpose of the visit was to encourage the kids, who were from a low-income area, to read more. Excited, I made sure that my schedule was clear for the day so that I could follow the team around the classrooms and visit with a few of them.

Over the course of the two hours that they were on campus, however, my excitement turned to concern and even worry. I had not begun my work with the human brain, so I was not operating with a brain science lens, but as I observed the football players in the different classes, I could tell that several of them were having difficulty reading to the class.

Their last stop was a second-grade class, where one of the veteran players (who had been in the NFL for more than fifteen years) was asked to read a Dr. Seuss book. He had difficulty pronouncing several words. Several of the second graders began correcting him, and the adults in the room shared looks of concern. At one point, one of the more rambunctious six-year-olds (who clearly had some issues with underactivity in his prefrontal cortex) blurted out, "Hey, mister! You don't know how to read, do you?" The rest of the class began to chuckle. The NFL player stopped reading, looked at the kid, smiled and then replied, "Yeah, I have more trouble reading these days. I think I've been hit in the head one too many times."

Brain injuries matter, and they often result in lifelong difficulties and struggles. Our brains were not meant to be placed in a helmet and then slammed against other people's helmets. They were not created to have a soccer ball slammed against them at seventy-five miles per hour. They were also not meant to be used as a punching bag in boxing, kick-boxing or ultimate fighting, otherwise known as mixed martial arts.

Psychological Trauma and Stress

Many people are not aware of the impact that psychological trauma can have on the brain. People who experience significant emotional events can be left with debilitating struggles, painful memories and disturbing emotions that leave them feeling depressed, anxious, hopeless, helpless and more.

In fact, there is growing evidence that the change in brain function in individuals who have endured significant psychological trauma is shockingly similar to that in people who have suffered physical trauma. One example of this is military personnel who experience emotional/psychological trauma on the battlefield, only to develop post-traumatic stress disorder later. (We touched on PTSD in Chapter 6, in the discussion of conditions that affect the basal ganglia.)

Individuals with PTSD typically start developing symptoms within three months of a traumatic event. In young people, typical events that can cause PTSD include experiencing or witnessing physical abuse or sexual abuse, sustaining a severe injury, witnessing domestic violence in the home, being involved in a serious accident and witnessing or experiencing an act of violence in the school or community.

Symptoms of PTSD include intrusive memories, flashbacks, dreams and reliving the traumatic event for minutes or even days at a time. Individuals with PTSD might develop increased anxiety, irritability, anger, guilt, shame, self-destructive behavior, hallucinations and trouble sleeping, and they may become easily startled or frightened. Additionally, they might have difficulty with concentration, memory and relationships. A general feeling of emotional numbness and a lack of interest in activities might occur. Often, it takes a great deal of effort for those suffering from PTSD to avoid thinking or talking about the traumatic event.

People who experience significant emotional/psychological trauma face a host of difficulties in their lives. Not least among these is a constant high level of stress. As we experience stress, our body releases a hormone called cortisol as a way to help us combat the stress and feel more balanced and calm. Unfortunately, consistently high levels of cortisol in the brain have now been shown to cause physical damage to brain cells.

You should try to minimize stress throughout your life. If you have experienced a significantly emotional experience that has caused you a great deal of pain and struggle, it is in your best interest to seek help in order to ensure that your brain stays healthy and functional. See Chapter 17 for information about stress-relieving techniques for relaxing your brain.

Environmental Toxins

In the following brain scans of one individual, what stands out? Clearly, you can see overall decreased blood flow and activity throughout the brain. On the surface, this brain looks quite similar to the brains of people who have used significant amounts of alcohol and drugs. This person, however, had not used any drugs, and he had barely consumed any alcohol. So what do you think caused this amount of damage to the most important organ in his body?

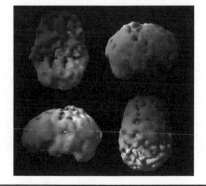

FIGURE 8.1: **ENVIRONMENTAL TOXINS**

As it turns out, this person had been using inhalants on a daily basis for years. These inhalants were not recreational drugs, however. Instead, he encountered inhalants on the job. He was a furniture finisher, a job that required him to spend his days working in a warehouse with stains and chemicals that were potent and quite toxic. To make matters worse, there was no air circulation in the warehouse, and he didn't think to wear a protective mask or even have a window or door open as he was working. He had been destroying his brain without realizing it.

Unfortunately, many children, teenagers and young adults use inhalants today. Sometimes, their exposure to inhalants is unintentional. Have you ever developed a headache after being around paint fumes? Other times, they purposely seek out inhalants to experience a cheap and easy high. Examples of inhalants include glue, felt-tip marker fluids, paint thinners, gasoline, butane in lighters, propane in propane tanks, nitrous oxide in whipped cream dispensers, refrigerants, helium, spray paint, spray deodorant, hair spray, cooking oil spray, static cling spray and even room deodorizer.

Some of the worst brain scans I have ever seen have been of individuals who used inhalants as their drug of choice. The chemical vapors can be absorbed by parts of the brain, as well as the nervous system. Exposure to inhalants can lead to asphyxia, suffocation, muscle weakness, depression, loss of smell, loss of hearing, nosebleeds, brain damage, death by cardiac arrest, suicide and much more.

What does this mean for you? You need to be more aware and to take precautions if you are working in situations where there are chemicals and vapors. If you can smell a chemical, that's your cue to ensure proper ventilation in the area or to leave the area until it is safe. This is especially true if you are painting, but it's also true in situations where you encounter air pollution and other environmental toxins, such as lead and other heavy metals, which have been shown to poison the brain.

Oxygen Deprivation

Oxygen is one of the few things that we absolutely need to have on a constant basis in order to live. Our brain is incredibly dependent on oxygen. In fact, that three-pound supercomputer housed in your skull, which accounts for just 2 percent or so of your body weight, actually consumes about 20 percent of the oxygen that you take in. It uses this oxygen to provide energy for its cells so they can transmit electrochemical impulses.

Anytime there is a disruption in the supply of oxygen, trouble arises. When oxygen stops flowing to the brain, brain cells start to die within minutes, resulting in a permanent loss of neurological function. This can manifest itself in a variety of ways, from lifelong mental disability to difficulty with concentration and memory. Oxygen deprivation in the brain can even result in death.

One troubling example of voluntary oxygen deprivation is the "choking game," which has increased in popularity over the years. Other names for this "game" include blackout, flatlining, space monkey, airplaning and space cowboy. Typically, kids engage in the "choking game." They get together and strangle one another with belts or even their bare hands in order to get a quick and temporary high. Unfortunately, this "game" comes with an extremely high risk of coma, stroke, bleeding into the brain, brain damage and death, due to the strain placed on blood vessels and the lack of oxygen to the brain.

Sleep Deprivation

Many teenagers and young adults pride themselves on going to sleep very late. And sleep is commonly missed for the sake of finishing assignments, earning money in jobs and maintaining relationships and friendships. Think of how many young people pull all-nighters in order to cram for a final exam. The irony is that sleep has been directly linked to how we process and remember information: the less we sleep, the more difficulty we have processing information, learning and remembering.

For young people, sleep is especially important. During sleep, our body releases growth hormones and also works to repair any damage that may have been done to it during the day. Also, during sleep our brain grows and develops critically important brain synapses (connections) between nerve cells, which help our brain run more efficiently and productively.

If you do not get enough sleep in the years before the age of twenty-five, there are a host of consequences that you might face. In the short term, these include decreased performance, memory impairment, decreased alertness, stress in relationships, an increased risk of bodily injury and cognitive impairment. In the long term, chronic sleep deprivation has been linked with ADHD, a poor quality of life, high blood pressure, stroke, heart attack and heart failure, depression, anxiety, mental impairment, obesity and a whole lot more.

Cigarettes

We have long been told that cigarettes are damaging to our lungs, and that smoking greatly enhances our chances of developing cancer and dying. We have all seen the pictures of the blackened lungs that have suffered through years of cigarette abuse. In some parts of the country, public service advertisements on television feature people who have lost fingers, toes or even a voice box after years of smoking.

Cigarettes have also been linked to damage in the brain. Smokers have been found to experience a sharp decline in the performance of their brain, to the extent that they are not able to perform as well on tests as those who have not used tobacco. Smokers also have higher incidences of high blood pressure and obesity, both of which have also been found to be damaging to the brain.

In addition, the prevailing view is that smoking causes about 40 percent of all strokes in individuals under the age of sixty-five. A stroke occurs when blood flow to the brain is disrupted by a clogged artery or a broken blood vessel. Sadly, when a stroke occurs in the brain, the parts of the brain affected typically die, meaning that the abilities controlled by those areas are gone.

Technology

The teenagers and young adults of today are living in an era that has changed the way the brain works more than any other in history. In large part, this is a result of the digital revolution: technology has been created that has forever changed the way we learn, communicate and operate in the world.

With technological devices, however, comes the need for instantaneous gratification. Today everything comes at us at a fast pace. We have instant communication, messaging and videoconferencing with other people. We get immediate news flashes through Twitter, which allow us to know what is happening before the news can report it.

This dependence on technology comes with some startling statistics. Nowadays teenagers and young adults spend almost eight hours of each day engaged in pursuits that hinge on technology, interacting with televisions, computers, music devices, mobile devices, video games and so on. *Those individuals who spend the most time interacting with technological devices each day have been found to have lower grades, a lower IQ and inferior cognitive power.*

Children who watch programs, movies or videos on television, on computers or on tablets, for example, have a vastly increased risk of

ADHD (a 10 percent increase for every hour of television watched per day). An increase in technology use among children has also been linked to poorer writing skills, lower levels of creativity, and a greater need for stimulation and activities that increase dopamine in the brain. Young people who are heavy cell phone users have been shown to make more mistakes in tasks that require memory, to have shorter attention spans and to do less well in overall learning.

Multitasking

When you sit down to do some work, how many technological devices do you operate at the same time? The developing brain today has grown accustomed to perching at the computer and having Microsoft Word, iTunes, email, instant messaging, Facebook, YouTube and an Internet browser open at the same time. At the workplace, people find themselves answering the phone, writing an email, checking social media, responding to text messages and working on projects at the same time. The worst part about this is that this is a point of pride for many of us. When confronted about doing so many things at the same time, we brag that our ability to "multitask" is what allows us to accomplish many important things at the same time.

But this just isn't true. Multitasking does *not* allow us to be more productive and efficient. Instead, it actually makes us more distracted, inefficient and unproductive. To those of you with active cingulates, I know that I have probably infuriated you. Allow me to explain.

While it seems logical that multitasking saves us time, research indicates that the exact opposite occurs. It seems that when we divide the brain's attention and focus between two or more tasks, there is a cost. Research has consistently concluded that if the brain tries to do more than one thing at a time, the result is a decrease in performance.

This is nearly impossible for people energized by multitasking to accept. This is especially true when you consider the dopamine fixes we get when we are constantly inundated with immediate responses. But as much as young people like to think that they are different and that today's technology has made people more efficient than in the past, the evidence still holds true: your brain operates best when it is allowed to focus on one thing at a time.

One exception to this is when people listen to music when they are focused on another activity. Performing one task while listening to music has been shown to increase overall speed and accuracy. This is because music activates a completely different part of the brain than what is used when completing tasks. In other words, the brain is not competing with itself for attention. For some people, the added stimulation of music can actually help them focus on a task at hand.

Other Examples of How Young People Are Damaging Their Brains

In case you hadn't realized this until now, teenagers and young adults are at an especially high risk of engaging in behaviors that show a lack of judgment and forethought. This is partly due to a lack of development in the prefrontal cortex, which is the part of the brain that keeps us from doing stupid things, and partly due to the deep limbic system, which makes us act out of emotion (rather than reason) and the desire to be accepted by our peers. Below are some examples of the dangerous things that young people do that often result in brain injury.

- **Car surfing:** This is climbing on top of a car and holding on while a friend drives. Yes, this gives you an adrenaline rush, but head injuries are almost guaranteed when things go wrong.

- **Drinking hand sanitizer:** When young people become desperate to consume alcohol, many turn to drinking cough syrup and/or hand sanitizer. This behavior has led to a high number of cases of blindness, memory loss and extreme agitation, and has been accompanied by a host of other dangerous and damaging side effects.

- **Vodka eyeballing:** This is when teens pour a shot of vodka directly into their eye sockets, which produces a rapid buzz as the alcohol enters the bloodstream directly through the veins of the eyeball. This has resulted in intense burning sensations, scarring, permanent damage to vision and the eyes, and even blindness.

- **Alcohol-soaked tampons:** This is the practice of soaking a tampon in vodka and then inserting it into the vagina or the rectum in order to become intoxicated. This can result in bacterial infections, yeast infections, the cracking and burning of vaginal/rectal tissue, alcohol poisoning and even death (due to how quickly the alcohol enters the bloodstream).

When Dr. Amen and I created the first version of our high school curriculum, we were asked to pilot it at a continuation high school in Orange County, California. Like at most continuation/alternative high schools, the student body was composed of kids who had been kicked out of traditional high schools due to drug use, alcohol use, violence, extreme behavior problems and/or pregnancy. The principal at the school was a fan of Dr. Amen's work and decided to make our course a requirement. From the first day of the course, the students' defiance was dialed higher than usual.

The course material was presented to the students very much the way the information is organized in this book. We taught the brain basics, covered many of the brain systems, discussed how they were related to behavior and decisions, and then empowered the students with the skills, knowledge and strategies to make changes in their brains and lives.

Soon after our first class, we noticed that these particularly troubled, problematic students became intrigued and engaged in the course. They had realized that our purpose was *not* to preach to them about the bad decisions they were making. Once they understood that our purpose was to teach them about the brain and how to optimize it, they started to *want* to learn more. Within a couple of weeks, other students from the school tried to crash our course because they had heard about the practical, relevant and important information we were teaching. Within a month, the parents of the students were asking if they could learn about the brain, as well.

One of the most memorable moments we had that semester was during the session when we talked about the different ways people hurt their brain. We discussed how drugs, alcohol, stress, a lack of exercise, sleep deprivation, negative thinking, smoking, excessive caffeine, too much television, violent video games, poor diet, environmental toxins, brain injuries, sports, traumatizing experiences, infections, oxygen deprivation and dehydration all worked to harm our brain.

As we went through the list, a wave of discontent and anger swept through the room. Students began shaking their heads, arguing with their neighbors and throwing their hands up in protest. It was obvious that this topic did not resonate well with the class, since many of the students engaged in several of these activities on a daily basis. While they had been fine with the material they had covered in the class up to that point, as soon as they felt like they needed to change their behaviors, things got ugly.

Finally, one student, who had clearly heard enough, shouted, "So how can you have any fun in your life?" We had heard this question often during our time teaching teenagers and young adults about the brain, so we were prepared. We led the students in an exercise we called "Who has more fun?" For the next fifteen minutes or so, we took the time to allow the students to contribute to a list of events and circumstances that they felt brought a great deal of joy to their lives. As you can imagine, this list included things like having a good career, making money, getting married, having kids, playing sports, participating in outdoor activities, socializing with friends, buying a nice car, traveling, having a good relationship with family, being involved with a church, living in a nice house and being successful overall.

After the students finished this menu for the ideal life of joy, we asked them to look at the brain scans of two teenagers. The first one showed the brain of a healthy and successful freshman in college who had obviously taken good care of his brain and body. His brain looked fantastic: it was even, symmetrical and well-balanced. Since the students were already familiar with brain scans and knew that what they revealed was related to behaviors and success, a few students even oohed and aahed in appreciation. The second brain scan was that of a senior in high school who had been using alcohol and marijuana recreationally (and only on occasion) throughout his high school years. His diet was poor, he rarely exercised and he didn't take care of himself. His brain had obvious areas of low brain function, a result of his poor care.

Finally, we asked the students to think about which of the two individuals had the best chance of achieving each of the joys in life that they themselves had listed. Begrudgingly, they agreed that it was the healthy student who had the best chance of having a fun life.

"Ah, hell," one student said. "We get what you are saying. Trying to have all the fun in life now only gives us fun for right now, but that means that we can screw up the chance of having fun later. But why does it got to be so hard?"

SUMMARY

- Our brain is soft and fragile, and our skull is hard.

- Our brain is not meant to be slammed into anything!

- Stress is disastrous for the brain.

- There are many environmental toxins around us that can damage the brain.

- The brain needs oxygen, sleep and proper nutrition.

- Technology can overstimulate the brain.

- Multitasking is good in theory but has been repeatedly shown to be ineffective for getting things done.

- Young people are at a higher risk of damaging their brains because the prefrontal cortex is not yet fully developed.

QUESTIONS

1. Which type of brain injury covered in this chapter did you find most surprising?

2. Why do you think people who play sports argue against there being any potential for brain damage/injury related to the sport?

3. Why do you think people continue to play sports that have been shown to be damaging to the brain?

9 WHEN THE BRAIN STRUGGLES

My first memory is of a time when I was almost four years old. I was sitting in the front passenger seat of my dad's truck, feeling confused. When I looked ahead through the front window, I could see a line of parked police cars blocking the road. I could clearly see several officers behind their car doors, aiming their guns in our direction. When I looked up through the sunroof, I could see a police helicopter hovering above us. When I looked to the left, I could see my mother waving her middle finger in the air and cursing at the police officer who was banging on the glass and yelling at her to open the car door. I was overwhelmed by the flashing lights, the wailing sirens, the deafening roar of the helicopter, and the screaming and yelling by my mother and the police officers.

Then, through all this chaos, I remember hearing a knock on the window to my right. I turned my head to look, and there in front of me, on the other side of the glass, was a young, pretty female police officer, smiling at me. Even more confused, I gave a half smile back and turned back to my mom, who was clearly distracted by the police officers at her window.

continued on page 120

Again, I heard a tap on the window next to me. When I turned to look this time, the officer was holding a lollipop in her hand. Now my interest was piqued. I looked at the lollipop, then at her. She was still giving me that pleasant smile, letting me know that she was here to help me and take care of me.

Then she looked at the car lock, then back at me. Then she looked at the lollipop and then back at me. She pointed at the car lock and motioned for me to pull it up. She brought the lollipop closer to the window and smiled even brighter. Even at the age of three, I understood clearly what was going on here. If I pulled up on the car lock, I could have a lollipop.

I looked back at my mom, who was now waving both of her middle fingers in the air and screaming with a ferocity I had never witnessed before. I turned back to the police officer, who was still smiling at me and holding the lollipop. I pulled the door lock up.

The last thing I remember is the door flying open and my being whisked out of the car at lightning speed. Yes, I sold out my mom for a lollipop.

Growing up in a family that suffered brain struggles was difficult, but not unbearable. Mostly, this was because I didn't know any different. Mental instability was the norm around me, but I didn't know what I was dealing with. My father was über-controlling about every aspect of his life, as well as the lives of those around him. His prefrontal cortex was underactive, which meant that he could be fairly impulsive and disorganized, and had difficulty following through. He dropped out of school when he was a junior in high school, and he never attended college. At home, our house was a collection of great ideas and projects that went unfinished. He would start a new project with the greatest of intentions, but then he would get bored or distracted by a new idea and would move on to another project before he completed the first one.

My father's cingulate (the brain's gearshift) was overactive, which meant that he was stubborn, oppositional, defiant and obsessive.

Arguing with my father was futile and exhausting, as he could spend an eternity proving why he was right and everyone else was wrong. As a teenager, I learned that if I wanted him to do something, I would have to present it as if I thought he wouldn't want to do it. This way, his oppositional nature would kick in, and he would end up doing what I wanted him to do in the first place.

Finally, my father's basal ganglia were overactive, which explained why he was anxious, was always predicting the worst outcomes, avoided social situations and was overly concerned about what people thought of him.

In the early 1980s, when my parents were young, my father was obsessed with the idea that my mother was cheating on him. He figured out a way to record all her phone conversations at home and spent countless hours fighting with her and accusing her of infidelity. After a while, my mother, who was already battling brain struggles of her own, was traumatized by my father's actions and accusations and began to get paranoid.

When my mother was pregnant with me, my parents often took road trips to visit his sister, who lived a few hours away. Sometimes she was stricken with bouts of morning sickness while they were on the road, and she would plead with my father to pull over so she could vomit. My father would ridicule her, fight with her and tell her that it was all in her head. He had my mother convinced that something was wrong with her mentally.

After years of living like this, my mother began to become paranoid in other areas of her life. She became convinced that people at work were conspiring to get her fired. My father would laugh at her and tell her that she was being crazy and paranoid, all while he continued to record her phone conversations and follow her to work to make sure that she was where she said she was going to be.

Finally, when I was about three years old, my mother snapped. She began hallucinating, and her paranoia increased exponentially, accompanied by delusions of grandeur and irrational thoughts and

continued on page 122

behaviors. From that point on, life in our household was consistently chaotic. My parents divorced when I was eight, and for years I dealt with abuse at their hands. My father was emotionally and psychologically abusive. My mother was psychologically and physically abusive. My father ultimately committed suicide days after my twenty-third birthday. Today my mother still struggles with symptoms of schizophrenia.

Given how complex, fragile and unique our brains are, it would be foolish to assume that we can go through life without struggling at some point. The fact is that more than sixty million Americans are struggling with something related to their brain at any given point in time. The following are three truths about brain struggles that we should be aware of in our own lives and in the lives of others.

Truth #1: Brain Struggles Are Real

When a person has chest pain, the typical protocol is to go to a doctor, answer a few questions and undergo some tests that look at your heart and lungs to ascertain what might be going on. These tests can include an echocardiogram, a CT scan, an MRI, blood tests and chest x-rays. Doctors try to arm themselves with as much information as possible in order to ensure their patients stay healthy and functional or to restore them to health.

When a person goes in to see their doctor about a mental health problem (ADHD, depression, anxiety and so on), the process looks very different. If you see a general practitioner, you might talk with the doctor about your symptoms only for a few minutes before he writes you a prescription for a fairly powerful medication that is often accompanied by some serious side effects.

The question is, Why don't doctors look at the organ they are treating? Too often, I have encountered people who dismiss brain struggles as being purely psychological, suggesting there is no underlying physical problem. Essentially, they argue that people can *choose* not to have ADHD, depression or anxiety if they really want to.

This is how my family viewed brain struggles when I was growing up. My sister always had difficulty with attention, focus, concentration and organization when she was in school, and she was prone to procrastinate. Typical comments on her report cards included, "Needs to apply herself," "Missing/late assignments," and "I would like to see more effort." Even when my sister actually did her homework, she would lose it or forget to turn it in. And when I tried to tell my dad that it might help her to look into the possibility that she had ADHD, he would get upset and say, "That's a bunch of crap! She's flunking school because she's lazy! There's no such thing as ADHD!"

From the moment Phillip Jr. sat in my office, I could tell that he didn't want to be there. Phillip was about sixteen years old and had been dragged to the Amen Clinics by his mother. Seated to Phillip Jr.'s right was Phillip Sr., who clearly did not want to be there, either. To the left was Phillip's mom, who looked like she was ready to pull out her hair from frustration and exhaustion.

During the history-taking portion of the interview, Phillip's mom told me she was convinced that Phillip had ADHD. She gave examples of how Phillip was impulsive, easily distracted and disorganized. While she was going into detail about her son, Phillip Sr. grew visibly irritated. He grunted, crossed his arms, blurted out snide remarks and interrupted the conversation to say that Phillip Jr. was just like how he had been growing up, and he had turned out just fine. Phillip Sr. quickly made it known that he believed ADHD was not a real disorder, but an excuse people gave for being lazy and unmotivated.

continued on page 124

About ten minutes into the interview, I stopped typing my clinical history report and just watched Phillip Jr.'s mom and dad go at each other. One would start to say something about Phillip Jr., and then the other would get upset and would interrupt, which would then cause the other person to lash out and interrupt right back. After a few minutes, I looked at Phillip Jr., who had clearly gone to his happy place. Though he was looking right at me, I could see that he had tuned out the entire conversation.

At that point I raised my hands to get the parents' attention, and then I asked Phillip Jr., "What do you think about all of this?"

"I know I must have ADD or something," he said. "I can't concentrate at school, and I lose things constantly. But I also know that it must have come from somewhere." And with that, he glared at both his parents.

"What exactly are you hinting at, son?" Phillip Sr. asked, upset.

"Oh, please," his mother interrupted. "Lord knows you are more ADHD than a kitten on crack. Just the other day, you—"

"I do *not* have ADD," Phillip Sr. interrupted. "If anyone has ADD, it would be you! You're the one who can't pay any of the bills on time—"

"Stop it!" Phillip Jr. screamed. "This is exactly the reason why I said I wouldn't come here unless we all get scanned. We are one ridiculously dysfunctional family!"

At this point, I let out a small chuckle.

"What exactly are you laughing at?" Phillip Sr. asked, glaring at me.

"I apologize. I just need to clarify something before we move on. Phillip Jr., you agreed to come here for an evaluation *only* if your parents got evaluated, too?" I asked.

"Hell, yes. I'm like this for a reason! Do you see what I have to put up with?"

"You can say whatever you want, but I know that my brain is as smooth and perfect as a baby's butt," Phillip Sr. said.

At that moment, I asked them if they wanted to get a sneak peek at their brain scans. They had been scanned earlier in the day. They all agreed, and I left the office for a few minutes to retrieve the pictures from the imaging room.

When I returned, I gave them a ten-minute explanation of what the brain scans would show, and I explained the different parts of the brain that we would be looking at, namely, the prefrontal cortex, the cingulate, the basal ganglia and the deep limbic system. As I talked about the prefrontal cortex in particular, I could see a look of concern come over all three of their faces. Clearly, many of the behaviors that are associated with the prefrontal cortex were ones that they had difficulty with. I also explained how genetics played a large role in how our brains are wired.

Then I showed them their scans. As you can imagine, all three of their brains had areas of decreased blood flow and activity in the prefrontal cortex. Impulsively, Phillip Jr. pointed to his father's brain scan and started laughing. Without any forethought, Phillip Sr. then smacked his son across the back of his head. Phillip's mother didn't realize what was going on, because she had been distracted by a text message she had received.

After a few minutes, however, Phillip Sr. said, "I hear what you are saying to me. I do. Everything that you said is related to this pre-functional court date part of my brain is everything I know I struggle with. But even though I hear you saying that I am the way I am because of my brain, I still can't believe it. Why is it so hard for me to accept this, even though I am staring at my brain right here in front of me?"

"Well," I said, "maybe we should talk about the cingulate."

Phillip's story is one of many that I accumulated during my time at the Amen Clinics. As a clinical historian there, my heart would often ache as I sat with patients and heard their stories about struggle, hardship, sadness, fear and desperation. For many, coming to our clinic and having their brain scanned was their last hope. They needed to see if there was any chance for healing and joy in their lives.

In each one of my clinical history appointments, I would allow time to talk about the brain as it related to human behavior. I would show patients the differences between scans of healthy brains and scans indicating that parts of the brain were underactive and/or overactive. Then we would take a look at their brain scan to see what it indicated. My job wasn't to make a clinical diagnosis or offer treatment suggestions (that was done by the psychiatrist in the final evaluation). Instead, my job was to offer patients a glimmer of hope by showing them that their problems were, indeed, real, that their symptoms and problematic behaviors were a result of brain function. For example, people who had been struggling with ADHD could see clearly that their prefrontal cortex was not functioning well. Some with depression and mood disorders were amazed to see how the deep limbic part of their brain was lit up like a lightbulb. Those struggling with anxiety found peace when they saw that the basal ganglia part of their brain was overactive. They were also relieved that there was brain function of any kind, since they had been predicting the worst-case scenario all along. Those with OCD would become obsessed with understanding every aspect of the cingulate and what they needed to do next.

For most of the patients who walked into my office, this was the first time in their lives that they had learned critical, yet practical, information about the brain. This was the first time that they were able to connect the brain to behaviors and mental disorders. Then, as they grasped how their brain had been working against them, something powerful would happen. They realized and accepted that they weren't bad people. This entire time they had been struggling with a brain that wasn't working right. Then something even more magical would happen. These people would realize that there was hope. With the proper treatment plan, they could change the way their brain worked.

When we accept the truth about the brain and accept mental illness for what it is, we have the responsibility to fight against the destructive stigmas attached to mental health. For far too long, people with debilitating mental disorders have been discriminated against, marginalized and judged. These illnesses are deemed somehow less legitimate than diabetes, cancer or asthma because of where the root of the problem is.

For the most part, this is because mental illness is still widely misunderstood. How much of the brain did you truly understand before you started reading this book? Did you see the connection between the different brain systems and certain behaviors? Were you aware of just how complex and fragile the human brain is, and how it is ultimately responsible for your level of joy and success in life? Unfortunately, most people are not. The exciting part is that this is changing.

Truth #2: Brain Struggles Are Complex

Once we have accepted the idea that brain struggles are very real and prevalent in our lives, we must also consider another important truth— that brain struggles are complex. You already know that the brain is regarded as the most complex organ in the known universe. Yet when we think about mental disorders, such as ADHD, depression, anxiety, OCD and so on, we tend to view them as single and simple disorders. Doctors often want to work from a checklist, diagnosing a problem, prescribing a medication and then sending you on your merry way. The truth is that this is a dangerous mind-set to have when you consider all that is happening with the patient.

Amy was twenty years old when she first went to her doctor with a mental complaint. For some time, she had felt like she had ADD. She had difficulty concentrating and focusing on tasks, and she was often frustrated that she was not achieving her full potential. When she sat with her doctor, who was a general practitioner, she told him how she was having trouble with attention, organization, focus and concentration, and how she had been procrastinating. She told him that her performance at work and in school had been suffering as a result, and that she knew it was time to get some help. After about fifteen minutes, Amy's doctor diagnosed her with ADHD, wrote her a prescription for a popular stimulant medication and said he wanted to see her in about six weeks.

The next morning Amy took her first dose of her newly prescribed stimulant medication and proceeded to have the day from hell. She described it as feeling like she was on methamphetamine or cocaine. She became hypersensitive, emotional, panicked and combative. Terrified, she called her doctor, who told her to cut the dose in half the next day. Even on the reduced dose, Amy responded the same way. Her heart began racing, and she began to feel paranoid and worried that she was going to die. After those two days of misery, Amy decided that she no longer wanted any help. She stopped taking the medicine and continued to struggle with her initial problems.

About five years later, Amy was at home, watching television with her newborn daughter, when she came across a special about the brain that was aired on public television. The show was *Change Your Brain, Change Your Life,* and she became fascinated by what she heard Dr. Amen discuss. Soon after, she picked up the phone and made an appointment at the clinic.

During her clinical history appointment, she began talking about her problems with focus, attention and organization, and about her proclivity to procrastinate. The clinical historian began asking her questions about depression and anxiety, which caught her off guard. She started to realize that she also struggled with motivation, poor eating/sleeping habits, low energy, feelings of hopelessness/ helplessness, nervousness and a constant worry about being judged by her loved ones. Amy concluded that she may have been feeling depressed and anxious because of her struggles with ADHD, that her inability to concentrate and focus had led her to feel anxious and depressed.

When she met with her treating physician at our clinic, however, she was shown pictures of her brain that suggested otherwise. Her brain's prefrontal cortex was not underactive, as is commonly seen in patients with ADHD. Instead, as you can see in Amy's brain scan, the deep limbic and basal ganglia parts of her brain were overactive, which is consistently observed in patients with anxiety and depression.

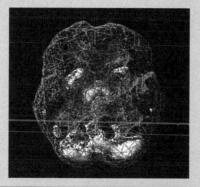

FIGURE 9.1: **AMY'S BRAIN**

When Amy fully grasped the meaning of her brain scans, she was dumbfounded. "Well, no wonder my stimulant medication made me go crazy! I was taking a medicine that was making my brain work harder when it was already working too hard!"

Amy was definitely struggling with focus, attention, organization and procrastination, but it wasn't because she had ADHD. Instead, it was because Amy was battling anxiety and depression. If you are nervous, anxious, depressed, lethargic and unmotivated, are not sleeping well and are feeling hopeless, do you think you would be able to concentrate and be organized?

Cases like Amy's are common in our clinics for one reason: the brain is complicated. Many people have come to our clinic thinking that they are struggling with anxiety, when, in fact, they are struggling with anxiety, depression and PTSD. Others feel like they are depressed, but the depression stems from having struggled with ADHD their entire lives. Most commonly, however, people don't realize that they are battling a number of different issues at the same time.

As much as we would like to find a simple solution to a problem, we have to accept the fact that it can never be that easy. The struggles that we face are often a result of how our brains work, and this functioning itself is the result of a vast network of trillions and trillions of cells and synapses that are all connected and communicating to one another inside our brain. For this reason, it is time to honor the brain and its complexity by being more thoughtful about how we treat people who are facing some very real and scary problems.

Truth #3: Everyone Struggles, but Not Everyone Knows When to Seek Help

When people finally accept the idea that a perfect brain is a myth, they respond in a multitude of ways. For some, a wave of relief and peace sets in as they realize that they are not alone in having a brain that might have some issues. For others, this understanding of the brain and how it relates to mental health offers a new glimmer of hope in the face of a history of depression and insecurity. Still, some people just gloss over this information and carry on as they were. Whatever the case may be, this truth remains: everyone struggles. The difference is the degree to which these struggles impact our lives.

When we face a difficult event in our lives, it is natural for us to go through a period of sadness and depression. Think of when a loved one passes away, for example. Grief is a natural part of healing as our brain processes the realities and difficulties of life. (In fact, if a loved one dies and someone close to him or her shows no emotional response whatsoever, that reaction might indicate a host of additional concerns.) During times of adversity, the deep limbic system might become more active, which makes things appear a little worse in the short term. People typically find healing by processing their grief with loved ones. Others might journal and find healing in solitude. The bottom line is that typically, it is okay to experience periods of feeling blue.

If depression begins to impede your ability to function, however, then the matter is much more concerning. If you feel symptoms of depression (such as sadness, changes in weight/appetite/sleep, hopelessness, helplessness, worthlessness, guilt, thoughts of death or suicide, loss of energy and little or no enjoyment from activities that were once pleasurable) for more than two weeks, then it might be time to seek help from a professional. If left untreated, symptoms of depression can persist for years and can cause significant changes in personality, work habits and relationships. In severe cases, depression makes it nearly impossible for a person to eat, sleep or get out of bed.

It can be difficult for those afflicted with depression to seek help, because they feel helpless, hopeless and/or worthless. They are sad and unmotivated to do much of anything. The thought of healing and feeling better is impossible thanks to negative thoughts. Because they do nothing to get better, they continue to feel bad. It is a dangerous and self-fulfilling prophecy that robs people of their joy and their chances for healing.

We can all relate to symptoms that stem from the prefrontal cortex, including having difficulty with focus, concentration, organization, learning from mistakes, forethought and judgment; exhibiting impulsivity; and having a tendency to procrastinate. In times of stress, these symptoms are oftentimes compounded.

For many people, however, such symptoms, which are related to ADHD, are more than just personality quirks and they are not limited to periods that are particularly difficult or stressful. For some, these symptoms are so prevalent that they prevent individuals from achieving success in school, at work, in relationships and in life. When individuals have symptoms of inattention, impulsivity and/or hyperactivity that have persisted for at least six months and have caused impairment in social, academic or occupational functioning, then help should be sought. It should be noted that typically, some of these symptoms have to be present before the age of seven years old in order for a patient to be diagnosed with ADHD. For individuals whose symptoms are severe enough to cause impairment in their lives, we often look to see if there is any evidence of brain injury or toxicity, as well.

Helping individuals with a struggling prefrontal cortex can be a challenge. Patients might find it difficult to stick to a treatment plan because they forget to do so. Or they might abandon the treatment plan when they start to feel better because they feel they are healed. We have to remind them that they must continue with their treatment plan in order to keep their brain working right. When a person with poor vision wears prescription glasses, they see better. This doesn't mean that they are cured and can stop wearing glasses. The glasses are what allow them to see better, so they need to keep wearing them.

This same philosophy holds for many of the struggles that we know are associated with the brain. For instance, many of us have some obsessive-compulsive tendencies that manifest themselves in interesting ways. I, for one, am known on campus for having the most OCD-like office of any professor. The books on my shelves are perfectly aligned, and they are organized by height and author. I have specific lighting to provide an ambience that is calming and inviting. I also do not allow clutter to gather on my desk or in my drawers. Does this mean that I have a psychiatric condition that needs intervention? Nope. But if my desire to have my office in perfect condition became an obsession, and I felt as though something terrible would occur if something were out of place, then I'd know it was time to seek help.

I also know that my basal ganglia are overactive, so I have a tendency to become anxious and to predict the worst in situations. At times, I have found this to be of benefit. I tend to work extra hard on projects to ensure that my work is of the highest quality. Other times, I have found this to be a problem (such as when I fly or have to sample the cuisine in other countries). If these worries became constant or obsessive, and if they lasted for several months, affecting my ability to feel joy, then I would know it was something I should talk to a professional about.

What Does It All Mean?

At this point, you should realize that having a brain struggle should not elicit any sense of fear, stigma or shame. We all have moments when things don't work quite like they should.

What often separates people who succeed from those who fail is how they react when things go wrong. When I was twenty-three years old and faced with the emotional trauma surrounding my father's death, I recognized that I needed help from a trained professional. I knew that the healing and processing that needed to be done were beyond my own ability. I did not feel ashamed or victimized by what had happened. I wanted to get through it and move on.

I have often worked with individuals who cannot get past the stigma associated with brain struggles. Some are too embarrassed or ashamed to tell others the thoughts, struggles and hurts that they have been facing. Others are stubborn and feel like they should be strong enough to handle things on their own. Those who are depressed don't see the point in seeking help, because their depression makes them feel hopeless and helpless. I have also worked with people who build up iron defense walls that are rooted in anxiety and a fear of being judged.

I often feel sad for these people, because they are the only ones standing in their way. I can see how they are hurting. Their brains are working against them, yet they refuse to step outside of their comfort zone in order to seek healing and hope for a brighter tomorrow. If there is one thing that I have learned about working with people who are struggling, it is that they are the only ones who can get themselves on the path toward healing. Nobody can force it upon them. They have to want it.

If you find yourself struggling with feelings, thoughts, emotions, behaviors or tendencies that are causing you a great deal of discomfort, pain, shame or embarrassment, it is critical that you know that you are not alone. More important, if you find that your struggles are serious enough that they are impeding your joy and detracting from your ability to succeed, then you must know that it doesn't have to stay this way. As complicated as the brain is, the exciting news in all of this is that we have the power to do something about it.

SUMMARY

- Brain struggles are very real, surprisingly common and biological in nature.

- Brain struggles are complex, which means that single, simple explanations for them are often unhelpful.

- We all have our struggles, but it's important to know when to seek help.

QUESTIONS

1. How has your opinion of mental illness changed as you have been reading this book?

2. How has your insight and knowledge about yourself changed thus far?

3. What else would you like to know about the brain as you seek to change your brain and your life?

10 GETTING TO KNOW OTHER PEOPLE'S BRAINS

You've read a lot about the brain, and you understand how integral it is to your personality and how you function on a day-to-day basis. Now it's time to look outside of yourself and think about other people's brains. The ability to succeed and find joy in life is not just dependent on you and your brain. Our lives are filled with interactions and relationships with other people. Growing up, we rely on our parents to raise us in a home full of love, support and guidance. We rely on teachers to teach us the information and skills necessary for us to become functional and independent members of society. We rely on coaches, pastors, mentors and other adults to encourage us to take risks, to point out the potential consequences of our actions and to act as the prefrontal cortex for us as ours continue to develop.

We also rely on friends to make us feel loved, connected and happy, and to encourage us when we are going through difficult times. We rely on girlfriends and boyfriends to give us purpose and allow us to be vulnerable, open and completely honest as we try to find our way in life. Indeed, much of who we are and how our brains function is a result of the relationships and interactions we have with other people. These relationships and interactions even have the power to change the way our own brains work!

If much of our joy and success is dependent on other people, this means that we are at the mercy of their brain function. What happens if the people in our lives are living with struggling brains of their own? Although I had had my share of hardship dealing with adults in my life who had brain dysfunction, this idea didn't really resonate with me until I started training teachers about the brain.

As I worked with Dr. Amen at his clinics over the course of a decade, I was always excited whenever we held conferences, workshops, lectures and seminars across the country and around the world. In the early years I managed the book tables in the back, and it was always encouraging to see how passionate and thrilled people were as they learned about the brain and strategies to change it and improve their lives.

I soon realized how powerful this information on the brain could be for students in schools. Dr. Amen agreed, and our first step was to write a high school course that would teach students about the brain. We also developed videos and programs that detailed and explained the dangers of drugs and alcohol. We started to get teenagers all across the country and the world thinking about the brain.

Several years ago I developed a training seminar entitled *Teachers as Brain Changers*. The aim of this seminar was to educate teachers, principals and counselors about the brains of their students. My hope was that I could empower these educators with strategies and skills that would allow them to meet the needs of all the various learners in their classrooms. I started with some brain basics, gave some details about the developing brain and then covered some of the specific regions of the brain and how they related to human behavior and student success.

I devoted the remainder of the seminar to specific strategies that would help students reach their potential in the classroom.

I was dismayed when the seminar went miserably. For the first half of it, the teachers, principals and counselors were thoroughly engaged as they learned about the brain, its different parts and what each was responsible for, its complexity and its fragility. I felt proud of myself as I watched them vigorously writing down their notes, asking questions and connecting the material to their own lives. But then, as I tried to shift the focus to applying this knowledge to the students in their classrooms, things went south.

Those with an overactive cingulate were stuck on how this information related to them personally (some were offended and defensive). Those teachers, principals and counselors with overactive basal ganglia were now worried that they themselves were struggling with mental illness and wanted more details. Some with an overactive deep limbic system became hopeful when they realized that their symptoms of depression were highly treatable, and so they insisted on discussing the different ways they could optimize their own brain function. And, of course, those with an underactive prefrontal cortex were fine no matter what, because they had spent much of the morning talking to neighbors or playing with their phones and computers.

At first, I couldn't understand why these educators were unable to focus on how the information I had presented applied to students in the classroom. Then it suddenly dawned on me. I had been operating under two assumptions: first, that they had already developed a keen understanding and appreciation for the brain, and second, that they were somehow different from the rest of the population and were immune to any brain struggles.

This was the first time that the vast majority of the teachers, principals and counselors had been given information about the brain that was practical, relevant and meaningful to their lives. Behaviors and symptoms that many of them had been struggling with for years were put into context, and suddenly they started to understand who they were and

continued on page 140

how their brain was responsible for their joy, success and even their effectiveness in the classroom.

I quickly began to understand that certain things needed to take place before these educators were able to focus on the brains of others. First, they needed to understand the basics of brain function and development. Then they needed to spend time exploring and understanding how their own brain functioned and how that related to their experiences. They needed to learn about themselves and think through how their brain impacted their own lives before they could think about their students. Only after these two steps had been achieved were they able to focus their attention on understanding their students' brains and explore how this knowledge could result in a more engaging and productive classroom.

Teaching teachers and other adults about the brain was amazing for me because it allowed me to learn additional things about brain function and development. One of the key things I learned was that teenagers and young adults need to be aware of the brain they are relying on for their success. If the parents, teachers and other adults in a young person's life are all operating with brains that are not functioning well, success and joy will be that much more difficult for the young person to achieve. But once teenagers and young adults know about the brain and can understand why other people act the way they do, they can adapt their behaviors and build relationships with others in a way that is mutually beneficial.

In order to get to know other people's brains so that you can increase your own chances for better relationships and future success, you must go through a series of four steps.

Step 1: Understand the Basics of Brain Function

Although this will seem obvious by now, it still needs to be said. Think of the brain systems you've learned about and how they are connected with behaviors, symptoms and emotions. Remember how the prefrontal cortex is like the CEO of your life, in charge of forethought, judgment, focus, attention, impulse control, empathy, planning and organization. Think of how the cingulate is like the gearshift of your thoughts, making you stubborn, defiant, obsessive, compulsive and argumentative when it is overactive. Also, think of the deep limbic system and how it is highly active during the teenage years, which makes it difficult for teenagers and young adults to think with logic and not their emotions. And finally, recall the basal ganglia and how they are associated with anxiety, an expectation of the worst and irrational fears.

Understanding the basics of brain function is critical for a number of reasons. First, it allows you to better comprehend human behavior and emotions. For example, I once worked with a teacher at an elementary school who struggled with severe obsessive-compulsive disorder. One morning I arrived at work earlier than usual and decided to stay in my car and eat the breakfast burrito I had just bought. A couple of minutes later, this teacher arrived and parked her van on the street. She got out of the van, locked the doors and checked each door to see if it was locked before she activated the car alarm. After the alarm had been set, she went around the van and checked to make sure each door was locked again.

Confident that her van was secure, she walked across the street, toward the school. But then she suddenly stopped. She turned around and stared at her van for a few seconds. I could see a look of concern sweep over her face. She walked back across the street and checked all the door locks on her van. Then she deactivated her car alarm and unlocked all the doors. She opened and closed each door one by one. She locked the van once again and then proceeded to check to make sure each door was locked. Then she reactivated the car alarm and checked all the doors again.

Confident that her van was secure, she walked across the street and toward the school. But then she suddenly stopped. She turned around and stared at her van for a few seconds. I could see a look of concern sweep over her face.... She repeated the same OCD pattern for nearly thirty minutes. I know this because I sat in my car and watched her. At the time, I didn't know what to think. Part of me felt bad for her, but I was mostly confused and concerned. When I talked to other teachers about it, they let me know that she was well known for having severe OCD. Most of the staff avoided her, and almost all her students disliked her. Unfortunately, she was also a running joke for some of the staff on campus, who would often move her mailbox or items in her classroom around just to watch her freak out.

A few years later, after I had learned about the brain and had changed my outlook on the struggles related to brain function, I remembered this teacher. I thought about the pain and suffering that she must have endured on a daily basis. She didn't choose to be this way. She did not find joy in having OCD. She didn't understand what was happening, and she didn't seek out any help for her disorder.

When you develop a deeper understanding of the basics of brain function, you begin to see behaviors and problems for what they really are: issues in the brain. You start to feel more sympathy and have more empathy when you encounter the most difficult people. Instead of getting angry and being disrespectful to teachers who are horribly disorganized, cluttered or distracted, you can feel bad for them, as you understand the struggles they might be facing. Instead of becoming defensive and angry when you deal with parents, teachers or other adults who are oppositional, stubborn and inflexible, you can have compassion for them, since you understand that this might be how their brains are wired. You will also understand that becoming angry and defensive in return will most likely make things worse.

Having a deeper understanding of how the brain functions also makes you less judgmental of others. When a person is struggling with OCD, the comment they hear most from others is, "Just stop doing that and you'll be fine." Meanwhile, the person with OCD is thinking, *Oh, really? Just stop? Do you think if I was able to 'just stop,' I wouldn't have done that by now? Do you think I enjoy the fact that I can't live a normal life?*

People with depression and anxiety face similar judgments. When people are depressed, others are apt to tell them that they just need to stop feeling sad. They don't understand that the sufferer doesn't really have a say in the matter. The sufferer doesn't have control over the flood of thoughts that make him or her feel sad, hopeless, helpless, worthless and even suicidal. Similarly, when people tell those struggling with anxiety disorders to just stop worrying, the result is the same. These responses are futile, serving only to make sufferers feel worse about themselves and how they feel.

Step 2: Know Your Own Brain

Step 2 happens naturally. Once you've learned about the brain, the first thing you want to do is understand how it all relates to you. This is an important step. Knowing how your brain functions provides you the opportunity to work on improving it for the future.

As you consider how your brain might be wired, it is important to look at what you do each day that works to help or harm your brain. Ask yourself the following ten questions:

1. What am I eating each day to ensure that I am giving my brain the proper nutrition it needs to work well?

2. Am I giving my brain enough physical activity and exercise to ensure proper blood flow?

3. Am I putting my brain at risk of injury by playing sports or engaging in other physical activities that can cause it harm?

4. Am I introducing toxic substances (e.g., drugs, alcohol, inhalants) into my brain that can cause permanent damage?

5. Are there symptoms, behaviors or feelings that I am dealing with that are hindering my happiness and ability to function?

6. How often do I find myself being held hostage by negative thoughts and worries that come automatically and forcefully?

7. Which of my personality traits and characteristics can be considered strengths? How are these related to brain function?

8. Which of my personality traits and characteristics can be considered weaknesses? How are these related to brain function?

9. What can I do to change my brain in order to minimize any unwanted behaviors and/or feelings?

10. What can I do to optimize my brain so that I have the best chance at joy and success in my future?

After you have answered these questions, you will have a better understanding of the role that your brain plays in your life. You can reflect on which parts of your brain might be overactive or underactive, and then you can determine how this affects your performance in school and at work, and how this impacts your relationships with family members, friends and romantic partners. You can also judge how hard you have worked to protect the most important organ in your body, and then you can make plans to optimize your brain for your future.

If you are one of millions of teenagers and young adults who are facing one of the countless brain struggles that negatively impact happiness and chances for success, knowing how your brain is working against you is the first step in getting better. Once you have let go of the stigmas attached to mental illness and can accept the fact that your brain can change with the proper intervention, you can then take the steps toward a brighter tomorrow.

Step 3: Accept That You Cannot Change Others

Sometimes, as people begin learning about the brain and themselves, they want to convert everyone into a brain fanatic. Feeling empowered by the knowledge of how they can change their brain and their life, they work tirelessly to get others to change their brain and life, too. Some people are even convinced that they can change the way other people's brains work. Trying to change other people is an exercise in futility.

Let's be honest here. Some of you reading this book *right now* are doing so only because it is a class requirement. Someone in your school or school district may have been excited when he or she grasped the importance of learning about the brain and its impact on the lives of teenagers and young adults, and so the decision was made to make *all* students read this book. If this was the case, then perhaps you're rolling your eyes, yawning, disagreeing with the content and just waiting for it to end so that you can get your grade and move on with your life.

If this is how you feel, then consider for a moment how pointless it is when somebody tries to change the way *you* feel or think. No matter what that person may say to you, you cannot change yourself unless you *want* to do so. If you have an overactive cingulate, this becomes exponentially more difficult for you, since your brain is physically wired to make you argumentative and oppositional. Just as you think it's impossible for people to change you, you shouldn't try to change other people. Instead of investing so much time in imagining what others should have or could have done, just let them be.

Here are some examples of the different thoughts we have about trying to change other people:

- "If he would only apply himself and actually think about what he is doing, things would be so much better."

- "If she would just stop focusing on the negative, maybe she could be happy."

- "If he hadn't cheated on me, we would still be together and everything would be great."

- "If my teacher was a good teacher, then maybe I would have gotten a good grade in the class."

- "If my boss wasn't so stubborn and difficult, my job would be so much easier."

- "If my parents weren't so controlling and demanding, then maybe I would actually want to spend more time with them."

In each of these situations, people get stuck in a pattern of negativity for two reasons. First, they focus on other people, whom they have no control over. Second, they focus on aspects of other people that the people themselves might not be able to change. As ridiculous as it might sound, they cannot help thinking, *If only people did not behave like themselves.*

When people learn to accept that they have no control over how others behave, life becomes more peaceful and enjoyable. When they also accept the idea that others behave the way they do because they might be wired to do so, then stupid behaviors and mistakes are easier to forgive. Instead of wasting all your time and energy focusing on why people act the way they do, and then becoming frustrated, why not focus on what you can do about it?

Step 4: Adapt Your Behavior to Others' Needs

Now we get to the fun part! Once you have reshaped your behaviors in a way that is compatible with the brains of those you interact with, you will be amazed at how enjoyable life can be. This is particularly true when you consider the brains of parents, teachers and bosses.

One place to start is by scoring parents, teachers or other adults who play a large role in your life on the Brain Systems Quiz, the same quiz you took in Chapter 1. This will help you better understand how their behaviors might be related to specific regions of the brain. Of course, keep this confidential, using the results of the quiz only for your own understanding and benefit. You wouldn't want your parents, your teachers or other adults to be offended should they discover you've been trying to figure out their brain.

The next chapter is filled with strategies and examples of how you can adapt your behaviors to the brain function of others in a way that makes everybody's lives easier and more enjoyable.

SUMMARY

- Understanding the brain function of those around you is helpful as you navigate through relationships, classes, jobs and life's other opportunities.

- To understand how others think and behave, you first have to understand the basics of brain function.

- To understand how to interact with others, you first have to understand how your own brain works.

- You have to accept that you cannot change how another person thinks, acts, interacts or behaves.

- When you adapt your behavior to the needs of others, you will often derive more joy and achieve greater success.

QUESTIONS

1. Why is it important to know how other people's brains work?

2. Why do you think some people try so hard to change others' behaviors?

3. How has the information presented in this chapter made you think differently about the most difficult people in your life?

11 ADAPTING YOUR BRAIN TO THE BRAINS OF OTHERS

In Chapter 4, I told the story of my cousin and how his overactive cingulate was going to war with his parents' overactive cingulates. He was stubborn, but his parents were even more stubborn. And even though he knew that he would face severe consequences for his acts of defiance, he still continued them, because he couldn't keep his cingulate under control.

After he accepted the fact that he wasn't going to change his parents, he began to change the way he interacted with them. He realized that every time he argued with his parents, things got worse for him. When he was quiet and respectful during heated moments, things got better for him. Even when he knew that his parents were in the wrong, he learned to stay quiet and accept the fact that his parents needed to be right. Within weeks of altering his behavior toward his parents, he witnessed his relationship with them change dramatically and his life became more enjoyable and happier. There was no longer any fighting or arguing in the home. His parents began to loosen their reins on him, and he began to enjoy the increased freedom and flexibility.

One evening a few months later I went out to dinner with him and his younger sister. We talked about how drastically different the environment at home was. His younger sister had this take on it: "Well, duh. I learned a long time ago that when I was in trouble, all I needed to do was shut up, apologize and say that I had learned my lesson. My parents forgave me, and we moved on. Not my brother. He would sit there and fight back and argue. I'm glad you finally learned to shut up!"

Dealing with Prefrontal Cortex Issues

If you have a parent who is struggling with an underactive prefrontal cortex, it is important to know that he or she will always have some issues with impulsivity, empathy, organization, judgment, focus, planning and procrastination. If Mom or Dad forgot about your open house, sporting event, awards assembly or even graduation, it probably wasn't because they didn't love you or thought that something else was more important.

Parents who have an underactive prefrontal cortex might say things in the heat of the moment that they regret later. They may have a tendency to be impulsive, to give little thought to consequences. Such parents may have difficulty with follow-through or may make promises that they can't keep. If you have a parent like this, you already know this often leads to hurt feelings and resentment. But when you realize that their failures as parents are rooted in their brains, you can lay down the groundwork for forgiveness and healing, which will strengthen the bond between you. Don't test your parents to see if they remember an important event on their own; this is just setting them up for failure. If you want them to follow through on something, you must do your part by reminding them—in other words, you must act as their prefrontal cortex for them.

If you have a teacher who comes to class late every day, unprepared, stressed and overwhelmed, it can be particularly frustrating. Perhaps the teacher's office or desk looks like something right out of an episode of *Hoarders*—even if he or she claims to know where everything is among the piles of clutter. During class, students may be amazed and overwhelmed when the teacher veers away from the syllabus and goes off on a vast number of tangents in a single hour. It may seem as if the class is never on task or on schedule with the work or assignments, which becomes particularly challenging at test time, when students may feel like the teacher never covered the material. Perhaps the teacher fails to give much feedback on essays. Grades can be infuriating, because they may be late, inaccurate or impossible to figure out.

If you have a teacher like this, it is important for you to accept that you will need to work harder in this class. Because the teacher may not cover much of the material in class, it is your responsibility to read the textbook, the handouts and your class notes to ensure that you are well prepared for any tests that are administered and essays that are assigned. If you have questions, ask them in a way that is respectful and sincere. Remember, it's up to you to follow up on your discussions to ensure that your needs are being met.

Having a boss at work with an underactive prefrontal cortex can also be frustrating. Perhaps the boss is easily overwhelmed and reactive in times of stress. Bosses like this might say things impulsively that are rude or even disrespectful. Work assignments can be challenging if the boss doesn't set a schedule until the last minute. Hours on your paycheck may be wrong, and it may feel like there is always a crisis of some kind brewing.

In these situations, a couple of different things can occur. With bosses who are unorganized and easily overwhelmed, but still kindhearted and respectful, you can excel and advance quickly. If you are able to work hard and compensate for your boss's inefficiencies, people will take notice. But some bosses with an underactive prefrontal cortex are much more difficult to work with. These are the bosses who cannot work well, because of problems with executive function, planning, organization, procrastination and impulsivity. I've heard people describe working in this environment as living in hell. If this is the case, the best advice is, alas, find another job.

Dealing with Cingulate Issues

Parents with an overactive cingulate are fairly easy to spot. They demand that things always be a certain way. They often hold on to certain viewpoints or beliefs with extreme ferocity and dedication. If you made a mistake in the past, these parents will remind you about it until the day you die, making sure you feel guilty about it as often as possible. If you piss these parents off, you may hear about it for years and have to endure stories about what they would have done differently. And they will hold on to the grudge until the day that *they* die. Everything is a battle for them, and their children are often frustrated with the amount of control that these parents attempt to exert over them.

In dealing with such parents, the most important thing to realize is that arguing or fighting back will only *make things worse*. You have to learn to let things go, stay quiet (even if they are wrong) and maintain an attitude of respect and appreciation. Once such parents feel like they are not engaged in a power struggle with you, you might be surprised how quickly things can get better.

Reverse psychology is also an interesting strategy to use with these parents. The following figure illustrates different ways you might approach a conversation with a parent who has an overactive cingulate:

1a. "Hey, Mom, I was offered a job at the hospital after school. I think I want to take it because it might help me get into medical school."

OR

1b. "Hey, Mom, I was offered a job at the hospital after school. I told them I probably wouldn't be able to do it because I know you would be worried that it might affect my grades."

1a. "I don't think this is a good idea, sweetie. Getting a job will only distract you from your homework, and if you don't do your homework, you won't get into medical school."

OR

1b. "What?! Why would you think that? I think that this is a good idea because it might help you get into medical school."

2a. "Hey, Dad, there is a clearance sale over at the sporting goods store this weekend. We need to go and buy that ping-pong table you promised me for my birthday."

OR

2b. "Hey, Dad, I'm gonna go to the sporting goods store on the weekend and check out the ping-pong tables that they have on clearance. I think I have saved enough money to finally pick one out for us. I was going to ask you if you'd like to go, but I figured you can't because you are probably working."

2a. "Not gonna happen. I have too much work to do this weekend, and you know that we are tight on finances right now."

OR

2b. "Don't be stupid! Of course I can go with you, but only if we go later on in the afternoon tomorrow. Plus, let's see how much it costs. Didn't I promise you one for your birthday anyway?"

3a. "Hey, Mom, you need to make sure that you pick me up at exactly 3:00 p.m. today at school. You cannot make me late for my appointment again."

OR

3b. "Hey, Mom, I rescheduled my doctor's appointment for 4:30 instead of 3:30. I told them that this would be easier since you are always late getting out of work."

3a. "I'll be there when I have the chance to be there. Don't lecture me about me making you late. How many times have *you* made *me* late for work because you don't know how to wake up in the morning?"

OR

3b. "I am not *always* late getting out of work. How dare you blame me and embarrass me like that! You change that appointment back to 3:30 and I will see you at 3!"

FIGURE 11.1: **YOU VS. YOUR PARENTS**

In dealing with teachers who might have an overactive cingulate, things can get tricky. These teachers are rigid, inflexible, controlling, stubborn, argumentative and oppositional, and they may be obsessive-compulsive. Sometimes, they will be unfair in their grading, because of the expectations they have set. They are inflexible about changes to the syllabus, class policies, lessons and exams. Essentially, they have a system that works for them, and that's that.

Students who do not understand the brain often have the most difficulty with these types of teachers. This is especially true for students who also have an overactive cingulate. In these cases, you have stubborn vs. stubborn. In all cases, however, the result is the same: arguing with stubborn teachers only makes your life difficult.

Even if your teacher is awful, there is not much you can do about it. A teacher like this has a reputation that follows him or her. Some years brave students might band together and revolt against the teacher and petition the administration or the school district. The teacher may get a reprimand, but tenure keeps him or her secure. The next year nothing changes.

If your teacher is oppositional and stubborn, the best thing to do is work your hardest to stay on this teacher's good side. Do not challenge the system or the teacher. But remember, it isn't anything personal. It's the teacher's brain that makes it difficult for him or her to be flexible, consider various options and be comfortable in acknowledging mistakes.

In my college years, I had a professor who very much fit this profile. He was extremely stubborn and argumentative. On our first night of class, his greeting to us was, "Okay, listen to me. Some of you may not like me, how I teach or what I teach in class. That's perfectly fine with me. But if I want to spend the entire class talking about the different ways people pick their noses, you'll just have to sit back and listen to me. Don't bother trying to complain about me or get me fired, because the truth is that I am tenured, and I bring way too much money into this university with my research for them to think about firing me. So, let's get started, shall we?"

That first class, he spent the time talking about nuclear submarines that patrolled the coast of the Pacific Ocean. Every few minutes he would walk up and down the aisles of desks, and we students would get a whiff of alcohol. He also had a reputation for being intoxicated during most of his classes.

The next week our first assignment was due. It was a ten-page research-based literature review that had us focus on the impact of technology in education. The assignment was to be APA formatted and needed to address each of the points in his outlines. The week after, minutes before class ended, he passed back our papers and started cleaning up to leave.

Within seconds a wave of discontent and outrage filled the room. Nearly the entire class had received zeroes on the assignment. I had to look at my paper for about thirty seconds before I could process the fact that right at the top of the first page was a big red 0. And right below the title on my cover page was this: "Title too long." There was no additional feedback. It soon dawned on me that he hadn't read a single page of my paper. I had spent an entire week writing a ten-page research paper, and he had stopped reading at my cover page, giving me a zero because my title was too long.

As the professor was trying to make his escape, several students blocked the door, forming a line to talk with him. I got in line, as well, to make my case about how my efforts needed to be acknowledged, and how it was ridiculous to give me a zero because my title was too long.

As I waited my turn, I listened to how the other students made their cases, paying close attention to the professor's responses. One by one, the students argued their cases, and the professor flat-out rejected

continued on page 156

them. At one point he shouted, "I gave you instructions. I laid out my expectations. It's not my fault you couldn't follow directions."

Then one of my classmates went up to him to complain that he was given a zero because his title was too long. The professor looked at him blankly and said, "APA guidelines recommend that titles not be more than twelve words. You had more than twelve. I'm not going to waste my time reading your work if you can't even follow directions for your cover page." Dismayed, the student stomped away and slammed the door shut as he left.

I looked back down at my paper and began counting the number of words in my title. Thirteen. Hours of work had gone down the drain, and all because of one word on the cover page. My heart sank as I attempted to process the many emotions that were flooding my head.

"Can I help you?" my professor asked. It was my turn to complain about my grade, and I just stood there, frozen. In the few seconds it took me to respond (which felt more like an eternity in my head), I played out many possible scenarios about how this conversation could go. On the one hand, my activated cingulate wanted to protest my grade and make him see the amount of effort and energy I had put into the rest of the paper. My prefrontal cortex, however, fought right back, telling me that it would just make things worse.

"Oh, yes," I began, quickly trying to compose my thoughts. "I…just wanted to thank you for the clarification on APA guidelines." This caught my professor off guard, and his face shifted from defensive to unsure. I continued, "I appreciate that you are holding us to a higher standard, and I wanted to let you know that I will be much more thoughtful and careful with the rest of my assignments."

The professor just stood there, speechless. Clearly, he was unsure how to respond. After a few seconds (which, I'm sure, felt more like an eternity in *his* head), he said, "Uh, sure. No problem. What was your name again?"

I knew then that I would have to ensure that all my assignments for that class were modeled specifically to his needs and expectations.

There wasn't any room for creativity or individualism here. I needed to do exactly what he said. It wasn't my role to challenge him, because he was the professor, and I was not. And even though I wasn't going to learn a ton from him, I still needed a good grade in the class.

After I adapted my behaviors to his teaching style, things were much more peaceful for me in the class. I would sit back and chuckle to myself as I watched some of my peers throw tantrums each and every time they received their papers back. I shook my head when they would then try to argue their case, which just made the professor upset and agitated. And in the end, I was one of only five students who received an A in the class.

Dealing with bosses who have an overactive cingulate is not much different than dealing with parents or teachers with the same issue. At work, you will need to be *that much more* careful to avoid arguing with and upsetting a boss who is stubborn, oppositional and inflexible. Even if you don't like how your boss manages you and your coworkers, you work for your boss, and he or she has the power to terminate you.

This doesn't mean that you can't be creative, however. If you work hard to be agreeable and to stay on the boss's good side, he or she can become your best friend and supporter. When I worked in room service at a popular hotel in downtown San Diego, I had a manager who was extremely stubborn and argumentative. After I witnessed numerous employees upset him, I realized that I wouldn't be doing myself any favors if I tried to prove him wrong in any way. Instead, I agreed with him as much as I could and worked hard for him. At one point, when I wanted to have a full weekend off so that I could go to Six Flags with my friends and family, I knew it was going to be a hard sell. My manager was notorious for denying requests for days off. Several weeks prior to the trip, however, I waited for the perfect time to bring the issue up.

One day, after the morning rush had ended and we were sitting around talking, I found my moment. I asked him if he had plans to take any trips over the summer with friends or family. He mentioned the possibility of going out of town with his wife and kids, and so I asked him several questions about it, showing a high level of interest. After a few minutes, he asked me if I had any travel plans.

I responded, "Nah, not really. I mean, a bunch of my family members are going on a trip up to Six Flags pretty soon, but I'm not planning on going with them."

"Why not?" he asked.

"I would love to go, but I know it would be impossible to take time away from work."

"What do you mean?"

"They scheduled the trip for a weekend that I know is typically busy for us."

"Ahh. Yeah, that would be tough."

"I know! It made a few of my cousins upset. In fact, one of my cousins offered to pay for my ticket since my birthday is that week. Another cousin even offered to call you to try to convince you to let me have the time off."

He laughed, but with a look more of concern than enjoyment.

"I told them that work was more important than going to Six Flags, and that I didn't want to trouble you with having to find coverage for me during a busy weekend."

"Hmmm." At this point, I could sense that his mind was working, that he was considering possible scenarios.

"Really, it's okay, though," I said. "I don't want you to worry about it or think that I'm upset if I don't go. I know that you don't like to approve weekend vacation requests. I'll be fine."

"You know what? Let me think about this. I don't think we are supposed to be *that* busy that weekend. Plus, with the new person we just hired, I think she will have had enough training to justify you taking those couple of days off. But this means that you'll probably need to work the next couple weekends with no days off, to make things fair."

And with that, I got to enjoy my full weekend off with my friends and family at Six Flags a few weeks later.

Dealing with Deep Limbic System and Basal Ganglia Issues

It's a different story when you have to deal with people who are struggling with issues related to the deep limbic system and/or the basal ganglia. Their symptoms tend to be more internalized than externalized. Parents with deep limbic problems may feel depressed, sad, unmotivated and isolated. Teachers will be dry, boring and unengaged with the content or with their students. Friends and significant others will start spending more time alone and will let their relationships dissolve.

Too often we deal with others who are depressed in ways that aren't helpful to them. We might think that they just seem sad, but then we don't know what to do about it, so we just move on, letting them isolate themselves even more and fade away without much resistance. Other times, we try to help people who are depressed by telling them they need to stop being depressed. We try to force them to engage in fun activities, but when they resist, we oftentimes give up, assuming that they like feeling that way and don't want to change.

In these situations, however, the best thing to do is to acknowledge that they are struggling with a health condition, and then to play an active part in their journey toward healing. Always be there to listen to their struggles and emotions, and fight the urge to get frustrated and angry when your advice isn't enough to cure them. Help them understand the brain science behind their struggle, and encourage them as they think about seeking professional help. Most important, do not lose contact. Being a consistent source of support and love can mean the difference between life and death for a person who is struggling to find a reason to get out of bed in the morning and carry on with life.

When the important people in your life are struggling with an issue related to the basal ganglia region of the brain, helping them can be just as tricky and confusing. Let's consider parents. Parents are typically overly concerned about how their parenting skills are perceived, so they might have a tendency to be hard on their children in order to maintain the image that they are great parents. These parents may also have difficulty with socialization, so they prefer that their children stay at home as much as possible in order to avoid their own anxiety and nervousness that come with the children leaving the home. Other parents might be excessive worriers, frustrating their children with their fear of worst-case scenarios, even in the most harmless of situations. "No, you cannot go to the movies with your friends tonight! Your friend's car might break down, and you could be mugged by a group of thugs who are high on cocaine! They'll steal your cell phone and leave you stranded in the middle of the street!"

Friends and significant others with high levels of anxiety can be frustrating to deal with in a similar way. Boyfriends and girlfriends may become hypersensitive to criticism and feel hurt easily. Some may become obsessed with the idea that the other is cheating on them, or that they are doing something wrong in the relationship that is pushing the other person away. Sadly, it is the obsession itself that often becomes what pushes the other person away, as he or she tires of having to reassure someone with high anxiety that everything is fine.

When interacting with very anxious people, it is important to remain consistent and supportive at all times. When the fears are centered on the idea that you are upset with them for something that they might or might not have done, you can remind them that everything is fine. Research now shows that it is perfectly acceptable to call out the anxiety for what it is. In other words, it can be beneficial for you to help them see how their concerns are irrational and how they can talk back to their thoughts so that they can move on. You should let friends and loved ones know that they can always talk to you without fear of judgment. Always show empathy and understanding as you help them see that their fears, irrational thoughts and stress are the result of a highly active brain. Whatever you do, don't give up hope, and don't get frustrated. While anxiety is treatable, it does not change overnight.

What Does It All Mean?

By now, you can see why an understanding of the brain can empower you to be more happy and successful in life. Not only do you know more about yourself and why you are the way you are, but you also now have a better understanding of the people around you. When you let people be who they are and stop trying to change them to meet your needs, life becomes more harmonious and enjoyable.

However, when you realize that you can change the way *you* behave and think in order to meet the needs of others, amazing things will start to happen. You will notice that the people who have caused you the most amount of stress become your greatest allies. Your relationships with the people closest to you will become deeper and more meaningful as they sense that you truly understand and accept them for who they are, and without expecting anything in return. You may find that people start to gravitate toward you thanks to your ability to connect and sympathize with them.

Now it's time to get down to business. Part III is all about equipping you with very specific skills and strategies to optimize your brain and change your life.

SUMMARY

- People with prefrontal cortex issues do not choose to be impulsive, disorganized or forgetful.
- It is important to forgive and to help people with prefrontal cortex issues be successful.
- People with cingulate issues are often upset that they get stuck on thoughts, actions or behaviors.
- Arguing with people who have cingulate issues only makes things worse.
- People with deep limbic system or basal ganglia issues cannot help but feel depressed or anxious.
- It is important to empathize with people who struggle with deep limbic system or basal ganglia symptoms. Do not tell them to "get over it" or to "just stop" feeling sad, depressed or anxious.

QUESTIONS

1. How will you interact differently with people who might be battling prefrontal cortex issues?

2. How will you interact differently with people who might be battling cingulate issues?

3. How will you interact differently with people who might be battling deep limbic system or basal ganglia issues?

Part III

How to Change Your Brain and Change Your Life (Before 25)

12

HOW TO SHAPE THE YOUNG BRAIN

By now, your knowledge of the prefrontal cortex, the cingulate, the deep limbic system and the basal ganglia has made you the center of attention at parties and has drastically increased your social standing among your friends, teachers, coworkers and family members. What? No? Not really?

As silly as this may seem, it might be worth it for you to take a moment to appreciate the fact that you probably know more about the brain and how it relates to human behavior than your peers, teachers, coworkers and parents, many lawyers and even some doctors.

I've talked with at least a dozen young people who made an appointment with their doctor for help with ADHD, OCD, depression, anxiety and many other conditions after they learned that their brain was the reason for their unhappiness. One particular young lady told her doctor, "I think that I'm struggling with a combination of depression and anxiety. I'd like to know how you can help me calm down any overactivity in my deep limbic system and basal ganglia, because I'm pretty sure that's what's happening here." The doctor was quiet for a few minutes, trying to fathom how such a statement could have come from the mouth of an eighteen-year-old. Finally, he said, "Wait. I'm sorry. Can we start again?" Some doctors, unfortunately, become defensive and argumentative (think cingulate) when confronted with information that might be unfamiliar to them or different from what they know.

In 2011 my family and I moved to Oregon from California so that I could start a new job as a professor of education. We were excited, nervous and very stressed. Moving a thousand miles away from home with a two-year-old and a five-month-old is never fun.

The first three days after we moved were a complete nightmare. The home we had rented was in awful shape, and we were desperately trying to clean, sanitize and paint it (both my wife and I have obsessive-compulsive tendencies when it comes to living in a clean home). On the third night, I put our son to bed and crashed on my bed. About a half hour later, I woke up to my wife freaking out.

She told me that our daughter, Emmy, was making weird movements that looked a little like seizures. Panicked, I got up and went to go look at her. She was crying, but the movements had stopped. We stood there for a long while, trying to figure out what had just happened and what we should do about it. We decided that if it happened again, we would call the doctor.

The next day it happened again, and I was able to record it on my phone. Every few seconds, Emmy would grimace and her arms would shoot up alongside her body. As each spasm occurred, we could see the fear in her eyes, and we felt powerless and terrified.

The next few weeks were among the most frustrating and disheartening weeks of our lives. The ER doctor told us that she was fine. Even when we showed her the video I'd made, she said it looked like my daughter had gas or was perhaps "trying something new with her arms."

"I'm sorry, Doctor, but I know enough about the brain to know that what's happening here is neurological and not gastrointestinal," I replied.

Offended that I had challenged her, she told us that there was nothing she could do, and she suggested we go see a pediatrician.

The following day the pediatrician became gravely concerned when he saw the video. He said it reminded him of a condition called "infantile spasms," which is a rare but debilitating condition that needed to be addressed as quickly as possible to prevent brain damage. He referred us to a highly respected and well-known pediatric neurologist in Portland.

When we met with this doctor, it was clear that he was operating with a cingulate that was highly overactive. He had a very specific way of doing things, and it was obvious that he didn't appreciate being challenged or questioned in any way. When my father-in-law, Dr. Amen, wanted to talk with the doctor to help in any way he could, the pediatric neurologist became defensive, saying that we needed to trust his judgment. He told us to start Emmy on medication that cost ninety-six thousand dollars, was not covered by insurance, and came with a list of severe and debilitating side effects. We felt it was our right to do more research and look for alternative treatments. When the neurologist realized that we weren't going to take his advice blindly, he kindly declined to treat our daughter. Ultimately, we did find a good team of doctors who were open to working with us in researching all the options to help make our daughter better.

The point I am making here is that you really cannot depend on other people to know what's going on in your brain. Not many people understand how the different parts of the brain are related to specific behaviors and brain struggles. More important, they aren't the ones living in your head—you are.

Brain Systems Quiz

Back at the beginning of Chapter 1, you were asked to rate yourself on different bahaviors that relate to the brain. You were also asked to set aside your answers for use later on. Well, it's later on!

Before we get to the strategies for changing your brain and your life, I'd like you to take the quiz once again.

BRAIN SYSTEMS QUIZ

Please rate yourself on each of the behaviors listed below using the scale provided. If possible, have another person who knows you well (e.g., a parent, a significant other, a close friend) rate you, as well, in order to construct the most complete picture.

0	1	2	3	4
Never	Rarely	Occasionally	Frequently	Very Frequently

_____ 1. Failure to pay close attention to details; tendency to make careless mistakes

_____ 2. Trouble sustaining attention in routine situations (e.g., homework, chores)

_____ 3. Trouble listening

_____ 4. Failure to finish things; tendency to procrastinate

_____ 5. Poor time organization

_____ 6. Tendency to lose things

_____ 7. Tendency to be easily distracted

_____ 8. Poor planning skills and a lack of clear goals or forward thinking

_____ 9. Difficulty expressing empathy for others

_____ 10. Impulsiveness (saying or doing things without thinking first)

_____ 11. Excessive or senseless worrying

_____ 12. Upset when things do not go your way

_____ 13. Upset when things are out of place

_____ 14. Tendency to be oppositional or argumentative

_____ 15. Tendency to have repetitive negative thoughts

_____ 16. Tendency toward compulsive behaviors

_____ 17. Intense dislike for change

_____ 18. Tendency to hold on to grudges

_____ 19. Upset when things are not done a certain way

_____ 20. Tendency to say no without first thinking about a question

_____ 21. Frequent feelings of sadness or moodiness

_____ 22. Negativity

_____ 23. Decreased interest in things that are usually fun or pleasurable

_____ 24. Feelings of hopelessness about the future

_____ 25. Feelings of worthlessness, helplessness or powerlessness

_____ 26. Feelings of dissatisfaction or boredom

_____ 27. Crying spells

_____ 28. Sleep changes (too much or too little)

_____ 29. Appetite changes (too much or too little)

_____ 30. Chronic low self-esteem

_____ 31. Frequent feelings of nervousness or anxiety

_____ 32. Symptoms of heightened muscle tension

_____ 33. Tendency to predict the worst

_____ 34. Conflict avoidance

_____ 35. Excessive fear of being judged or scrutinized by others

_____ 36. Excessive motivation (e.g., can't stop working)

_____ 37. Tendency to freeze in anxiety-provoking situations

_____ 38. Shyness or timidity

_____ 39. Sensitivity to criticism

_____ 40. Fingernail biting or skin picking

ANSWER KEY

Questions 1–10 = Prefrontal cortex symptoms

Questions 11–20 = Cingulate symptoms

Questions 21–30 = Deep limbic system symptoms

Questions 31–40 = Basal ganglia symptoms

If you answered two to three questions related to a particular brain system with a 3 or 4, struggles in that part of the brain may be possible. If you answered four to five questions with a 3 or 4, problems in that brain system are probable. If you answered six or more questions with a 3 or 4, problems in that brain system are highly probable.

After you have completed this quiz once again, you should go back and compare your answers to your answers from Chapter 1. You may be shocked at what you discover.

If you have found that your results were worse the second time around, take a deep breath and relax. This is common. Over the years we have found that when students take the Brain Systems Quiz *before* they truly understand the brain, they typically rate themselves as not having very many issues—simply because they lack the insight you now have. If you are like most people, you probably rated your behaviors higher on the scale now than you did the first time you took the quiz. This isn't because you have suddenly spiraled into mental illness and are now struggling with myriad brain issues. Instead, you should be proud and excited to know that you are that much more insightful and knowledgeable about your own brain and how it relates to you.

Shaping the Young Brain

Fifteen- to twenty-five-year-olds are in prime condition to shape their brain for success. In the first twenty-five years, the brain is undergoing an immense amount of development, and therefore teenagers and young adults must be keenly aware of every decision they make. Evaluating your actions, interactions, thoughts and emotions through the lens of brain science will ensure that you move toward your goals with few hindrances. You already know about the many ways you can hurt your brain. Now we're going to talk about strategies for keeping your brain growing well and for enhancing its connections.

Brain Reserve

One of the most difficult aspects of teaching people about the brain is trying to account for the vast differences in how each of our brains functions and responds to the outside world. When I teach young people about shaping the brain for future success, many ask the following questions:

- How come one person can do drugs and seem okay while another person does the drug once and dies?

- How come some people play football for years and are fine, but others play for only a little while before they have too many head injuries?

- How is it that one person can go through a bad breakup and be fine, but another person obsesses about it and can't get it out of his or her mind?

This is where Dr. Amen's idea of brain reserve comes in handy. Years ago Dr. Amen developed the idea of the brain reserve as the "cushion of brain tissue we have to deal with the unexpected stresses that come our way." All of us have a different amount of this brain reserve, and this amount is determined in part by the nature of our mother's pregnancy. In other words, if your mother smoked, drank alcohol, used drugs or was under lots of stress when she was pregnant with you, chances are that she decreased the amount of brain reserve you have, even before you were born. But if your mother felt relaxed, exercised, took her prenatal vitamins and ate a healthy diet, she probably increased your brain reserve.

After we are born, we are continuously increasing or decreasing our brain reserve. If you fell down some stairs at some point in time, you probably decreased your brain reserve. If you were exposed to chronic stress at home, same thing. If you used alcohol or drugs as a teenager, you decreased more of your brain reserve. Essentially, the more harm you have done to your brain, the more likely it is that your brain reserve has suffered.

But wait! There is good news! It's never too late to start increasing your levels of brain reserve. That's what the following chapters are all about.

SUMMARY

- Not many people spend a great deal of time connecting brain function to behavior.
- The developing brain is in the perfect position to set itself up for a lifetime of success…or failure.
- We were all born with a unique brain. We all have a different brain reserve level.

QUESTIONS

1. How did your results from the Brain Systems Quiz change since you first took it at the beginning of Chapter 1?
2. Why is it pointless to compare ourselves to other people?
3. What excites you the most about what you have learned thus far?

13 KILL THE ANTs

Every single time you have a thought in your brain, it can cause physical reactions, releasing chemicals into your brain and the rest of your body that can change the way you think and feel.

Let's imagine for a moment that you were to sit down for thirty minutes and write a paper about every single bad thing that people have ever done to you throughout your entire life. You would summon up all the times people have hurt you, teased you, embarrassed you, taken advantage of you and rejected you. How do you think you would feel afterward? Chances are that you would feel pretty awful. This is because every bad, mad, sad, hopeless, helpless and negative thought you have releases chemicals that make you feel bad. You might notice your hands get colder, you might start to sweat more, your heart rate might increase and flatten, your breathing might become faster and more shallow, and your muscles might tense up. This is the power that bad thoughts can have over you.

Now imagine that you spent that half hour thinking of every single positive, loving, selfless, hopeful, kind and joyful person in your life. Think of how they have supported you, loved you and helped you through thick and thin. How do you think you would feel after a half hour of this? You'd be skipping up and down the streets and singing out loud to yourself with joy and love in your heart. Okay, that might be a bit dramatic, but you see my point. Whenever you have thoughts that are happy, positive, hopeful and loving, your brain and the rest of your body release chemicals that work to make you feel good. Your hands might feel warmer, you sweat less, your heart rate normalizes, your breathing slows down and deepens, and your muscles relax.

Do you find this hard to believe? This is where the beauty and magic of brain science come in. In a study at the Amen Clinics, we scanned the brain of a person as he focused on happy and positive thoughts. The next day we scanned this person's brain again, but this time we asked him to concentrate on negative and depressing thoughts. The two brain scans you see below show the results of this study. The brain scan on the left is from when our subject was thinking positively. The brain scan on the right is the same subject's brain, but from when he was thinking negatively. Notice the power of our thoughts! When our subject thought negatively, his brain stopped working as well.

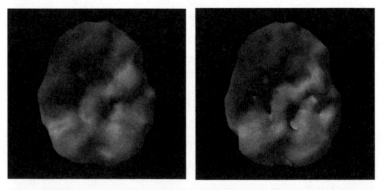

FIGURE 13.1: **HAPPY THOUGHTS**   FIGURE 13.2: **SAD THOUGHTS**

Automatic Negative Thoughts (ANTs)

"I know I'm going to fail the test."

"This homework is too hard."

"I can't ask her out, because she will say no."

"He wants to break up with me."

"The teacher never listens to what I say."

"My teacher hates me."

Do any of these thoughts sound familiar? These thoughts are only a tiny fraction of the many sad, negative and hurtful lies our brains tell us. When we allow these lies to resonate with us, they release chemicals that make us feel even worse.

For many years we have called these lies ANTs: **A**utomatic **N**egative **T**houghts. These are the thoughts that enter our brains and ruin our day. We do not ask them to invade our head. They just happen. In a way, they are like the black ants that invade our kitchens, picnics and barbecues. And as illogical, irrational and hurtful as our internal ANTs can be, we often let them go unchecked, which typically leads to our feeling upset, down and/or anxious.

The good news is you don't have to believe every stupid lie that goes through your head. Better yet, you can learn how to talk back to these lies before they take control. For many teenagers and young adults, this shouldn't be difficult to accomplish. Think of how good teenagers and young adults are at talking back to and challenging their parents.

In all our years of education, nobody ever teaches us how to think, what to think or what *not* to think. For many of us, there is a free-for-all going on in our minds, with random thoughts racing through without any rhyme or reason. Others of us get stuck on the same repetitive negative thoughts and can't get rid of them. Our negative thinking has taken control of our brains.

We need to fight to take control back. If you learn how to do this well, you will find yourself with improved brain function, enhanced moods, more success and better relationships. All that just from thinking about thinking!

The following are the nine different ANT species that steal our happiness and ruin our days:

1: The "All or Nothing" ANT

These are the thoughts that make you believe that things are either all good or all bad. If you get a 4.0 GPA one semester, you feel like you are the smartest person on the planet. If you get just one B, then you are a failure and the worst person ever. If you missed one question on the test, your life is over. If you stick to your exercise plan for a month, you think you are the best athlete ever. If you miss a day at the gym, you give up your gym membership and eat ice cream while watching television. The key is not to let yourself get caught up in thinking that everything is all or nothing. Master Obi-Wan Kenobi is a good person to remember when you think of the "All or Nothing" ANT. When Anakin Skywalker (before he turned to the dark side and became Darth Vader) was battling his "All or Nothing" ANTs, Obi-Wan cautioned, "Only a Sith deals in absolutes."

2: The "Always Thinking" ANT

These are the thoughts that contain overgeneralizations about situations. They typically start with words like *always, never, everyone, every time*. Examples of this ANT include:

"We never have any fun."

"You're always too busy for me."

"Everyone thinks I'm stupid."

"I will never feel happy."

My response when I hear utterances like these is, "Really? *Everyone* thinks you're stupid? There's not a single person on this planet who thinks you might have even a shred of intelligence?" My sarcasm is intentional in an effort to highlight the absurdity that typically accompanies the "Always Thinking" ANTs. These thoughts are destructive because they activate our deep limbic system and prevent us from seeing the many positives in a given situation.

3: The "Focusing on the Negative" ANT

This ANT is depressing because it makes you preoccupied with what's going wrong in a situation while ignoring the positive. If life gives you lemons, this ANT will definitely not make any lemonade. Instead, you think about how you don't like these particular lemons because they are too ripe and have a slight discoloration to them—even though you might actually love lemons. If you are given a glass that is completely full, this ANT will tell you that it is *too* full and that now you have to worry about spilling its contents. Even though you may have received a 98 percent on your final exam, this ANT will ruin your day because you missed two questions. Negative Nancy and Debbie Downer live with these ANTs, and it is important for us to combat them by trying to think of the positives in every single situation. For example, when my grandma passed away just a few weeks ago, I interrupted my sadness to appreciate that she had lived for ninety-five years, had had an incredibly full and active life, and was now free from the pain and suffering she endured during the last couple of years.

4: The "Thinking with Your Feelings" ANT

This ANT is typically based on events from the past. Examples of this ANT include:

"I feel dumb."

"I feel like I will never find love."

"I feel like you don't trust me."

These thoughts get you down, but you believe your negative feelings without even questioning them! When you have a "Thinking with Your Feelings" ANT, you should first look to see if there is any real evidence to support your feeling. You might want to ponder events from the past to see how they might be influencing your feelings now.

5: The "Guilt Beating" ANT

"Guilt Beating" ANTs are the thoughts that typically include words like *should, must, ought* or *have to*. Whenever we have these kinds of thoughts, we end up feeling guilty, which then triggers our deep limbic system. Teenagers and young adults already get enough guilt trips from their parents, don't you think? How many times do parents say things like, "You should call more often," "You have to be home for Christmas," or "You really ought to be nicer to him/her"? Our minds create our own set of guilt-beating thoughts, as well. One strategy to eradicate these ANTs is to replace the words *should, must, ought* or *have to* with new words and phrases, such as *want, would be helpful* and *it would be in my best interest to*. This might make you more motivated to get rid of the negative feelings behind your thoughts.

6: The "Labeling" ANT

"Labeling" ANTs are destructive because they attach a negative label to you or to others. When this happens, you banish the person you have just labeled to the group whose members include everyone else in your past with the same label, and then it becomes difficult or nearly impossible to deal with that person on an individual basis and in an appropriate manner. Examples of "Labeling" ANTs include:

"I'm stupid."

"You're a jerk."

"My teacher is arrogant."

"She is so irresponsible."

"He is mean."

When you feel an urge to use these labels, you have to stop yourself and clearly assess the situation. By doing so, you'll find that you can deal with yourself and others in a more reasonable way.

After my first year of high school, I decided I needed to be challenged more academically. I had received straight As my freshman year with little effort. Several of my friends were enrolled in honors and AP classes in our sophomore year, and, since my developing deep limbic system craved friendship, I made an appointment with my guidance counselor to see about switching to these classes.

After I made my case, she looked through my file for a minute and then shook her head. "I'm sorry, but I can't justify putting you into classes that I know you will fail in."

It took a few seconds for me to pick my jaw up off the floor. Shocked, confused and insulted, all I could muster up in response was, "Umm...what?"

"First off, you're Mexican, and both of your parents were high school dropouts and have never even been to college. Second, these classes are way out of your league. I'll just have to switch you back to the regular classes after the first week," she said. Hearing all the labels that she dropped on me had a devastating effect. I suddenly felt stupid. So I did the only thing I could think of: I got up and left.

When I got home, I called my dad and told him what had happened. He became enraged and hung up the phone after delivering a flurry of obscenities.

continued on page 182

About an hour later my guidance counselor called and said, "I spoke to your father. I will be switching your classes to the AP classes that you have requested, and I look forward to seeing you again in my office once you are failing them."

Having the active cingulate that I am wired with, you might imagine the kind of motivation that this provided for me. All I could think of was proving my guidance counselor wrong. This made me try even harder. At the end of the semester, I walked into her office with my 4.0 GPA report card. I sat down and gave her my report card while she continued typing on her computer. I had been looking forward to this moment for months! I was eager to hear her apology and to be told how impressed and happy she was for me.

Instead, she picked up the paper, glanced at it for about five seconds, placed it back on her desk and continued to type. Then, without ever making eye contact with me, she said, "Well, that's surprising." The next day, I was informed that I had been assigned a new guidance counselor.

7: The "Fortune-Telling" ANT

This ANT is very much a basal ganglia ANT; it has you predicting the worst possible outcome in any given situation. It's called the "Fortune-Telling" ANT because when you have such a thought, one would think that you have somehow acquired the ability to foretell the future. You seem to know what is going to happen. Your attitude and feelings are already negative, which makes you feel tense as you wait for the worst. Examples of this ANT include:

"I can't ask her out, because I know she's going to say no."

"What's the point of studying for my test if I already know I'm going to fail?"

"I'm going to get fired."

"My boyfriend is going to break up with me."

These ANTs are particularly dangerous because they can be self-fulfilling. If you have already decided that you aren't going to study for a test, because you think you will fail, you'll probably end up failing because you didn't study. Or if you actually do study, you may give up when you get to the more difficult parts of the material rather than buckle down and study harder.

8: The "Mind Reading" ANT

Similar to the "Fortune-Telling" ANT, the "Mind Reading" ANT is particularly dangerous because it requires you to assume you know something that you simply cannot know. Just as you can't foretell the future, you also don't have the ability to read another person's mind. With a "Mind Reading" ANT, you're convinced you know what somebody else is thinking, even though he or she has not told you, and you have not even asked. Examples of this ANT include:

"She doesn't like me."

"He's upset with me."

"My teacher hates me."

The best example I can give of a "Mind Reading" ANT is one I encountered back when I was an elementary school teacher. While my third-grade class worked on an independent assignment, I sat at my desk, getting prepared for the next activity. One of my students, Vicky, interrupted me. I could tell she was very upset. When I asked her what was wrong, she said that her best friend, Becca, hated her. When I asked why she thought this, she said it was because Becca kept giving her dirty looks and making faces at her.

When I looked up at Becca, I could tell that she was, indeed, making faces. To me, however, it looked more like she was in pain. I asked Vicky to sit down, and then I called Becca up to my desk. When

I asked Becca what was going on, she said, "Oh, Mr. Payne, I am so sorry. I have diarrhea really bad and am trying to hold it in!"

9: The "Blaming" ANT

The "Blaming" ANT is the worst of them all. Blaming happens when you hold someone or something responsible for the problems in your life. Blamers typically say the following:

"It wasn't my fault that…"

"If only you had done this…"

"It's your fault that…"

"How was I supposed to know that…"

These ANTs are harmful because when you blame something or someone else, you then become a victim of circumstance. You become powerless to change anything, and you do not hold yourself accountable for any of your own actions. When you realize your own role in the problems that are plaguing you, you can then consider how to change your actions and interactions to make things better.

Develop Your ANTeater

With swarms of these Automatic Negative Thoughts waiting to bring you down, it is important that you develop your own ANTeater to talk back to these nonsensical thoughts. Whenever you feel mad, sad, nervous or frustrated, write out your thoughts and then think about which ANT species they are. Then write down what your ANTeater would say to that ANT to kill it. Writing down the truth diffuses the negative feelings, and you will start to feel better. In Chapter 18, which contains the Two-Week Brain Smart Plan, you will be given an example of a chart that you can use to kill these ANTs and develop your own internal ANTeater.

The Work: A Simple but Effective Way to Kill ANTs

Another way to combat these Automatic Negative Thoughts is to utilize Byron Katie's approach, what she terms "The Work," which she details in her book *Loving What Is*. In this very wise book, Katie describes the amazing transformation that took place in her own life. At the age of forty-three, Katie, who had spent the previous ten years of her life in a downward spiral of rage, despair and suicidal depression, woke up one morning to discover that all those horrible emotions were gone. In their place were feelings of utter joy and happiness. Katie's great revelation, which came in 1986, was that it was not life that had made her feel depressed, angry, abandoned and despairing. Rather, it was her thoughts that had made her feel this way. This insight led Katie to the notion that our thoughts can just as easily make us feel happy, calm, connected and joyful.

This insight also led her to realize that our mind and our thoughts affect our bodies. "The body is never our problem. Our problem is always a thought that we innocently believe," she writes in her book *On Health, Sickness, and Death*. "Bodies don't crave, bodies don't want, bodies don't know, don't care, don't get hungry or thirsty. It is what the mind attaches—ice cream, alcohol, drugs, sex, money—that the body reflects. There are no physical addictions, only mental ones. Body follows mind. It doesn't have a choice."

Katie wanted to share her revelation about the power of thoughts with others to help them end their suffering. She developed a simple method of inquiry—The Work—which teaches how to identify and question deleterious thoughts and, in so doing, achieve a greater sense of peace, dominion, mental clarity and energy. The Work is simple. It consists of writing down any bothersome, worrisome or negative thoughts, then asking yourself four questions:

1. Is the negative thought true?

2. Can I absolutely know that it is true?

3. How do I react when I think that thought?

4. Who would I be without the thought? (Or how would I feel if I didn't have the thought?)

After you answer the four questions, you ask yourself whether the opposite of the negative thought is not true or is even truer. Then you apply this opposite thought to yourself, asking yourself, "How does the opposite thought apply to me personally?" And if the negative thought involves another person, you ask yourself how the opposite thought applies to that person.

Here is an example of how to use these four questions to kill the ANTs that are keeping you from achieving your goals. Mandy, who is seventeen, has a math test this week. Here is what she said:

Negative thought*: I will never pass the test.*

Question #1: Is it true that you will not pass the test?
"Yes," she said.

Question #2: Can you absolutely know that it is true that you will not pass the test?
Initially she said yes, because math was her worst subject. Then she thought about it and said, "Well, maybe if I studied really hard and met with a tutor, I could pass."

Question #3: How do you react when you have the thought *I will never pass the test?*
"I feel anxious, and I feel like a failure. I'm afraid that I will fail the class and have to take summer school, which means I won't be able to hang out with my friends and they won't like me anymore."

Question #4: Who would you be without the thought
I will never pass the test?

She thought about it for a moment, then said, "I wouldn't feel so nervous about the test, and I would probably feel better."

Turnaround: What thought is the opposite of "I will never pass the test"?

Mandy said, "I will pass the test." She added that she would pass the test if she studied hard. Then she felt a sense of empowerment, which encouraged her to study rather than give up.

What Does It All Mean?

On the surface, talking back to your negative thoughts might seem like common sense. It is incredible how often our thoughts lie to us, and how often we accept our thoughts as truth. We must realize that our thoughts come to us automatically. However, just because we have a thought in our head, it does not mean that it is true. When we believe these ANTs, they make us feel sad and down. But you have the power to take control of your thoughts.

When Dr. Amen and I started teaching high school students about the brain, one of the lessons that they appreciated the most was the one on ANTs. When we talked about the automaticity of our thoughts, and how they had the power to change our brains and make us feel bad, the students were all intrigued. They grew hopeful when they discovered that they didn't have to believe every ridiculous thought that they had, and that they could even learn how to talk back to the negative thoughts and take control. In the end, quite a few students commented, "Why hasn't anybody ever taught us about this before? Why don't our teachers learn about this stuff, too?"

- **All or nothing:** Thinking that things are either all good or all bad. If you stick to your exercise plan for a month, you think you are the best athlete ever. If you miss a day at the gym, you give up your gym membership and go back to being a couch potato.

- **Always thinking:** Overgeneralizing about a situation and usually starting thoughts with words like *always, never, everyone, every time.*

- **Focusing on the negative:** Preoccupying yourself with what's going wrong in a situation and ignoring the positive.

- **Thinking with your feelings:** Believing your negative feelings without ever questioning them.

- **Guilt beating:** Thinking with words like *should, must, ought* or *have to,* which produce feelings of guilt.

- **Labeling:** Attaching a negative label to yourself or others.

- **Fortune-telling:** Predicting the worst.

- **Mind reading:** Thinking you know what somebody else is thinking, even though he or she has not told you and you have not asked.

- **Blaming:** Blaming others for your problems.

QUESTIONS

1. What ANT do you feel you battle with the most? Why do you think this is?

2. What are the top five Automatic Negative Thoughts that you struggle with?

3. How has this chapter changed the way you feel about these Automatic Negative Thoughts?

14 EAT RIGHT TO THINK RIGHT

If you want your brain to work well, you have to give some thought to what you are feeding it. Eating right is one of the easiest and most effective strategies to improve your mood and to work more efficiently and productively at work and at school. With nutritious foods, you will feel better, have more energy and be mentally sharper.

Food Truth #1: You Really Are What You Eat

Throughout your lifetime, all the cells in your body, including your brain cells, replace themselves every few months. Your skin cells are on an even faster schedule, replacing themselves every thirty days. What fuels this amazing cell regeneration? The food you eat every day. Different foods make our bodies react and work differently. If you have a junk-food diet, you have a junk-food brain and a junk-food body. If you eat right, your brain works right, and you think right.

Food Truth #2: Food Is a Drug

You have probably noticed that what you eat during the day affects your mood and level of energy. Think of the way little kids bounce off the walls when they have sugary snacks like cake or candy. Or consider how a coffee addict acts before the first cup of the day. Food affects how we feel inside and out, and this is because food is a drug.

- **Food can make you feel worse.** If you scarf down doughnuts for breakfast, it won't be long before you start to feel foggy, spacey and slowed down.

- **Food can make you sleepy.** It's a cliché that the Thanksgiving turkey makes everyone sleepy after the annual holiday dinner, but have you ever noticed that just wolfing down a big lunch makes you feel like you need a nap?

- **Food can make you feel great.** Eating the right kinds of food will give you energy all day and help you focus.

During the writing of this book, I began to feel overwhelmed and fatigued at times. After a while, I realized that my ability to research and write was directly related to what I ate throughout the day.

I began to look into various meal replacement shakes that have been found to be healthy and nourishing to the brain and body. After some intense searching, I discovered an all-natural, organic raw powder that was free of just about every potential allergen and unhealthy substance (gluten, dairy, sweeteners and so on). In many of the reviews of the product, however, people noted how *awful* it tasted. One reviewer compared drinking the shake to "eating dirt from the ground of an organic farm." But many noted that if you were able to get past the taste, the shake would work wonders.

I gave it a try. Yes, it tasted awful. My eyes would water whenever I started to mix the raw powder with some coconut water and almond milk. After a few deep breaths, I would hold my nose and gulp the shake down. Then I would spend the next few minutes drinking some water, rinsing out my mouth and being as dramatic as possible.

Can you guess what happened? It worked! Within a few days, using this drink to nourish my brain and body in the morning and afternoon made me sharper, clearer and more articulate in my writing. I was more productive and focused, which saved me a great deal of time and energy throughout the writing process. Sure, the shake tasted gross, but in the end it was worth it. Please note that this is what worked for me; you might find something different that is better suited for you.

Food Truth #3: Diet Influences Everything in Your Life

Food doesn't just make the hunger pangs go away. It affects your physical health and well-being. When you eat a poor diet, your health suffers, your weight increases and your energy level decreases. Consuming nutrient-rich foods at every meal and at snack time, on the other hand, gives you a better chance of building a stronger immune system. Healthy diets are also associated with more stamina during physical activity, better weight control, more energy to get you through your day and an overall healthier appearance.

Your diet also affects your ability to think. Brain-friendly foods can rev up your mental sharpness and improve your chances of success in school or at work. Your diet can even have an effect on your relationships. When your diet makes you feel good, it's easier to be generous and pleasant to friends and family, and that makes everyone happier.

Food Truth #4: Our Diets Are Making Us One of the Fattest Nations on the Planet

Obesity is becoming an epidemic with a devastating impact on our health and our brains. Until recently, the United States had the highest level of obesity on Earth. Who took the top spot in 2013? Mexico. (Since I am both Mexican and American, you can imagine my reaction. *Lovely.*)

Obesity is determined by a person's body mass index (BMI), which is a ratio of their weight and height. Research from 2009 and 2010 indicates that more than one third of adults in the United States are obese. More than 50 percent of American women have a waistline greater than thirty-five inches, and 50 percent of American men sport waistlines of more than forty inches around. About one in fifty adults—about fifteen million people—are considered morbidly obese, which is defined as being at least one hundred pounds overweight.

Even more alarming, 34 percent of children and teens are currently overweight or at risk of becoming overweight, and 17 percent of children and adolescents are classified as obese. The number of children aged two to five who are considered obese has skyrocketed from 5 percent in the 1970s to 12.4 percent in the 2010s. That's an increase of nearly 150 percent. For the six-to-eleven age group, the number who are obese has increased from 6.5 percent to a shocking 17 percent, a whopping 300 percent increase, and for twelve- to nineteen-year-olds, the number who are considered obese has gone from 5 percent to 17.6 percent, more than a 300 percent increase.

Overweight children are more likely to remain overweight or become obese as adults, which puts them at greater risk for a variety of diseases and conditions that negatively affect brain function. Obesity has been linked to reduced cognitive ability, depression, decreased self-esteem and increased suicide attempts. Morbid obesity is associated with type 2 diabetes, heart disease and high blood pressure, as well as brain-related conditions, such as stroke, chronic headaches and sleep apnea.

There is much talk of rising health-care costs in this country. Obesity alone costs the U.S. health-care system up to $147 billion a year—that's $1,429 for each obese person.

10 Rules for Brain-Healthy Nutrition

The two brain scans below illustrate the impact of food on the brain. The scan on the left is of a young adult male who had consumed a healthy meal about an hour before. The next day he ate a fast-food meal that consisted of a double cheeseburger, a large fries and a milkshake an hour before he had his second brain scan, on the right. Notice the incredible difference in brain function between the two scans. This is the same person, and the scans were taken only twenty-four hours apart. The only difference was the food that was consumed. This is proof that what you eat matters!

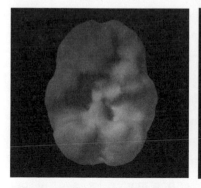 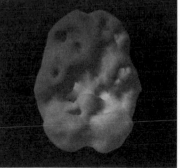

FIGURE 14.1: **HEALTHY DIET**   FIGURE 14.2: **POOR DIET**

The good news is that learning how to eat right to think right is not difficult. That shake I had aside, you don't have to hate everything you eat.

The following are ten easy rules to follow to ensure proper brain nutrition.

1. Maintain Adequate Hydration

Since your body is 70 percent water and your brain is 80 percent water, proper hydration is extremely important. Even slight dehydration increases the body's stress hormones, and over time, higher levels of stress hormones are associated with memory problems, obesity, wrinkles, irritability and difficulty thinking clearly. To avoid this, you should drink at least half your weight in ounces. So, if you weigh 150 pounds, you should drink 75 ounces of water each and every day.

Avoid caffeinated beverages, such as soda, energy drinks and coffee, and alcohol, all of which dehydrate you.

2. Watch Your Calories

Research shows that eating less helps you live longer. It's also a good way to control your weight and lower your risk for heart disease, cancer and stroke, since obesity is a major risk factor for these illnesses. Even better, managing calories in a healthy way triggers mechanisms in the body that increase the production of nerve growth factors, which are beneficial to your brain.

Teens and young adults tend to eat a lot of fast food. These days, many chain and fast-food restaurants post nutritional information about their food on their website or on the menu boards. So if you like to eat at Wendy's, Jack in the Box, Ruby Tuesday and other chain restaurants, you can go online to see how many calories are in your favorite meals. There's also a great website called Chowbaby.com (www.chowbaby.com) that features a calorie counter for foods served at a number of the most popular fast-food restaurants.

When Dr. Amen and I taught our first high school class about the human brain, we had the students complete the following exercise. In the school's computer lab, we asked the students to list the foods in a typical fast-food meal and then go to the restaurant's website and calculate the nutritional information for that. Then we instructed the students to find healthier meal alternatives at the same restaurants. The results were astounding.

	OMG No!	**Much Better**
Wendy's	Baconator (bacon cheeseburger)	large chili
	large fries	baked potato with sour cream and chives
	large Strawberry Frosty	water
	2,200 calories	*590 calories*
McDonald's	Double Quarter Pounder with Cheese	Chipotle BBQ Snack Wrap (Grilled)
		Premium Caesar Salad (low-fat balsamic dressing)
	large fries	Fruit 'N Yogurt Parfait
	large Coke	water
	1,550 calories	*550 calories*
Jack in the Box	Sirloin Cheeseburger	Chicken Fajita Pita
	onion rings	apple bites with caramel
	Fat-Free Mango Smoothie	water
	2,020 calories	*396 calories*

3. Increase Good Fats and Decrease Bad Fats

Try to incorporate into your diet more foods containing good fats, such as avocados, nuts, salmon, mackerel, olive oil, canola oil, peanut oil, safflower oil and corn oil. The good fats found in salmon, mackerel, and canola and soybean oil are high in omega-3 fatty acids, which are known as essential fatty acids because our body needs them. Most teens and young adults (and adults who are older than twenty-five, for that matter) don't get enough good fats in the foods they consume. You should try to eat one to two servings of fish a week, particularly fish like wild salmon (not farm-raised), which is high in omega-3 fatty acids.

In addition to adding foods with good fats to your diet, limit or eliminate saturated fats, which are found in red meat, eggs and full-fat dairy foods, such as whole milk and butter, and trans fats, which are found in store-bought cakes, crackers, cookies, potato chips and margarine.

4. Increase Good Carbs and Decrease Bad Carbs

Carbohydrates provide the fuel your body needs to perform physical activity. Some carbs are better than others. Complex carbs are the ones you want; they are found in fruits, vegetables, beans, legumes and whole grains. They take longer to digest and are chock-full of vitamins, minerals and fiber, all of which promote a healthy brain and body. The carbs to avoid are simple carbs, like table sugar, pastries, candy, sodas, fruit juices, doughnuts, white bread, pasta and white rice. Simple carbs are digested quickly, making you hungry again sooner, provide little or no nutritional value, and may promote disease and weight gain.

To see which carbs are high in sugar, check their glycemic index (GI) at www.glycemicindex.com, which rates carbs based on their effect on blood sugar levels. Low-glycemic carbs cause only small fluctuations in blood sugar levels, which helps you maintain energy throughout the day. High-glycemic carbs cause blood sugar levels to spike, then crash.

The key to good brain health is making sure most of the carbs you consume are low glycemic. Eating low-glycemic carbs that also contain a lot of fiber—such as vegetables, fruits, whole grains, beans and legumes—is even better for the brain. Dietary fiber can lower cholesterol, which promotes good blood flow. When choosing which fruits and vegetables to eat, go for non-starchy vegetables and low-sugar fruits, like broccoli instead of potatoes, and blueberries instead of pineapple.

Check out the following list for tasty low-glycemic, high-fiber foods you should stock up on. Some of them, like peanuts and cherries, make great snacks for the office or school.

Low-Glycemic, High-Fiber Foods That Taste Great		
Apples	Asparagus	Bananas
Barley	Black beans	Blueberries
Broccoli	Brown rice	Carrots
Cherries	Garbanzo beans	Kidney beans
Lentils	Low-fat yogurt	Oatmeal
Oranges	Peaches	Peanuts
Pears	Pinto beans	Plums
Prunes	Spinach	Yams

5. Ditch Artificial Sweeteners for Natural Sweeteners

Artificial sweeteners, such as aspartame, may be associated with arthritis, gastrointestinal problems, headaches, memory problems, neurological problems and a whole bunch of other health issues. Many people who suffered from migraines have reported that these headaches stopped after they gave up aspartame. Other people have noted that their confusion went away once they ditched artificial sweeteners. Still, others have reported that their extra weight started to come off once they stopped drinking diet sodas.

6. Eat Great Brain Foods

Foods that contain lots of antioxidants help your body and brain stay young. Several studies have concluded that a diet rich in antioxidants—they are found in many fruits and vegetables—significantly reduces the risk of developing cognitive impairment. For this reason, blueberries, which contain lots of antioxidants, have earned the nickname "brainberries" among neuroscientists. Eating a diet rich in blueberries reduces abdominal fat, lowers cholesterol, improves glucose levels and enhances your ability to learn new motor skills.

Keeping a stash of antioxidant snacks in your desk at work or in your backpack at school can help you steer clear of the candy bowl on the teacher's desk or the birthday cake in the office kitchen.

The 50 Best Brain Foods (in no particular order)		
Wild salmon	Tuna	Mackerel
Skinless chicken	Lean beef	Herring
Low-fat cheeses	Lean pork	Skinless turkey

Low-fat cottage cheese	Low-fat/ skim milk	Tofu and soy products
Low-fat sugar-free yogurt	Lentils	Pinto beans
Garbanzo beans	Almonds, raw	Raspberries
Black beans	Cranberries	Lemons
Walnuts	Grapefruit	Peaches
Blueberries	Red grapes	Broccoli
Blackberries	Beets	Red bell peppers
Oranges	Oats	Yams/sweet potatoes
Limes	Wheat germ	Olives
Cherries	Plums	Brussels sprouts
Kiwis	Spinach	Yellow bell peppers
Tomatoes	Whole wheat	Avocados
Olive oil	Water	Green tea
Black tea (decaffeinated)		Eggs (DHA-enriched are best)

The Best Antioxidant-Rich Fruits and Vegetables

Acai berries	Avocados	Beets
Blackberries	Blueberries	Broccoli
Brussels sprouts	Cherries	Cranberries
Kiwis	Oranges	Plums
Raspberries	Red bell peppers	Red grapes
Spinach		Strawberries

7. Balance the Foods You Eat

Your brain needs a balance of lean protein, complex carbohydrates and good fats. Try to include lean protein, such as skinless chicken or turkey, in every meal to balance your blood sugar levels. Adding lean protein to snacks and meals slows the absorption of simple carbs and helps prevent the brain fog that typically follows when you eat sugary snacks.

8. Limit Caffeine Intake

If you drink one or two normal-size cups of coffee or tea a day, that's probably okay. But any more than that can spell trouble for your brain and body. Here are five of the many reasons why:

- Caffeine restricts blood flow to the brain.

- Caffeine dehydrates the brain.

- Caffeine interferes with sleep.

- Caffeine is addictive.

- Caffeine can accelerate heart rate and raise blood pressure.

9. Reduce Salt and Increase Potassium Intake

A lot of people mistakenly blame salt for making them fat. Salt by itself does not cause weight gain, but it does cause your body to retain water temporarily, which can make it harder to zip up your jeans. The real issue is that salt is commonly found in large quantities in high-calorie processed foods, which you find in the snack aisle at the grocery store, at fast-food joints and at other restaurants. That's the likely reason why eating a diet loaded with high-salt foods makes you gain weight over time.

While you're cutting back on salt, it's important to increase your potassium intake. Studies have found that eating twice as much potassium as sodium can cut the risk of dying from heart disease in half and that taking potassium supplements lowers blood pressure. Foods that are high in potassium include bananas, spinach, honeydew melon, kiwis, lima beans, oranges, tomatoes and all meats.

10. Take a Daily Multivitamin and Fish Oil Supplement

The vast majority of Americans do not eat at least five servings of fruits and vegetables a day, the minimum required to ensure good nutrition. The American Medical Association now recommends a daily vitamin for everybody to help prevent chronic illness.

Fish oil is a great source for omega-3 fatty acids. The two most studied fish oils are eicosapentaenoic acid (EPA) and docosahexaenoic acid (DHA). DHA is a vital component of cell membranes, especially in the brain and retina. It is essential for normal brain development in fetuses and infants and for the maintenance of normal brain function throughout life. DHA appears to be a major factor in the fluidity and flexibility of brain cell membranes, and it could play a major role in how we think and feel.

Fish oil has been found to have many positive effects on mental and physical health. Canadian researchers have concluded that giving children omega-3 fatty acid supplements helps control ADD symptoms. Children had greater attention spans and other improved behavior when taking these supplements. For overfocused or obsessive ADD symptoms, add more DHA to your diet. For inattentive or low-energy ADD symptoms, add more EPA. Most people do best on a combination of both DHA and EPA.

Many studies have also shown that fish oil is beneficial for heart health, because it lowers triglyceride levels and has anti-inflammatory, antiarrhythmic, immune-enhancing and nerve cell–stabilizing properties. It also helps maintain normal blood flow and lowers the body's ability to form clots.

The Dangers of School Lunches

The National School Lunch Program (NSLP) and the School Breakfast Program (SBP), which provide subsidized meals for more than thirty million schoolchildren every day, continue to offer food that is too high in saturated fat and sodium and too low in nutrients, according to the third School Nutrition Dietary Assessment Study (SNDA-III), released in 2009. And the news gets worse. About 42 percent of the 398 public schools surveyed offered no fresh fruits or raw vegetables whatsoever with their meals. As a result, students who receive subsidized meals are not eating enough fruits, vegetables and whole grains. In addition, nonsubsidized cafeteria offerings and campus vending machines make sodas, candy, pastries, doughnuts, potato chips and other high-calorie, low-nutrient foods far too accessible.

By federal law, "foods of minimal nutritional value" are prohibited in school cafeterias during mealtimes. The problem is that the definition of "minimal nutritional value" hasn't been updated since 1979, when government officials were mostly concerned about making sure that foods provided certain amounts of nutrients. They set no standards or maximum allowed levels for calories, saturated fat or sodium, which leaves the door open for a host of junk foods.

One study (2012) looked at differences in the amount of junk food offered in elementary, middle and high schools. The researchers found that junk foods are more readily available in middle schools and high schools, with teens in high schools having the most access. In other words, your school food choices become less healthy as you get older.

SUMMARY

- Literally, you are what you eat.

- Food is a drug. It can make you feel good, bad, sleepy or energetic.

- Diet influences just about everything in your life.

- The average diet is not doing us any favors. We need to be proactive and thoughtful about what we put into our body.

- It is important to be hydrated, to watch your calories, avoid artificial sweeteners, limit caffeine and reduce your sodium intake.

QUESTIONS

1. How do you feel about your current diet?

2. Why do you think people eat so poorly?

3. What will you do to try to eat better?

15

BODY TIME TO BOOST THE MIND

In the Disney Pixar movie *Wall-E,* humans of the future are depicted as overweight blobs who eat all day and are transported in floating chairs wherever they need to go. Nobody so much as walks. Real-life humans of the present aren't much better. We drive to work, sit at a desk all day, use productivity tools designed to minimize the need for movement, and then drive home, where many of us spend the evening sitting on the couch, watching the tube or programs on our computers. We have almost completely eliminated movement from our day-to-day lives. This is bad news for our brains!

If you want a brain that is active, efficient and productive, you're going to have to get off your butt and move! Physical exercise acts like a natural wonder drug for the brain; exercising is the single most important thing you can do to enhance brain function. It improves the heart's ability to pump blood throughout the body, which increases blood flow to the brain. That supplies more oxygen, glucose and nutrients to the brain, which improves overall brain function. Research shows that exercise encourages the growth of new brain cells and enhances cognitive ability.

Due to budget cuts and an increasing emphasis on testing and test scores, many schools are cutting or eliminating physical education from the curriculum. This is terrible news for developing brains. Even if your school doesn't have a phys ed program, you still need to exercise. Find a local sports club, go for a run or gather some friends from the neighborhood for a regular game of baseball or basketball. Just move!

Physical Activity Improves Academic Performance

Many studies have been conducted that show a strong relationship between physical fitness and academic achievement, and a good number of books on this subject have been published. Dr. John J. Ratey's book *Spark: The Revolutionary New Science of Exercise and the Brain,* for example, describes an innovative physical education program at a school in Naperville, Illinois, that transformed the student body into some of the smartest kids in the country. For years, U.S. students have lagged far behind students from other countries in math and science. But in 1999 eighth graders at the Naperville school ranked first in the world in science and sixth in math.

So what was their secret? The school's phys ed program sidelined traditional sports in favor of high-intensity aerobic activity—a brief warm-up, a one-mile run and a cooldown. The only rule was that students had to keep their average heart rate above 185 bpm for the mile-long run, which means they had to run hard. Clearly, these bursts of activity paid off.

Here are some more studies to back up the connection between exercise and brain power:

- In 2005 the California Department of Education released a study that showed that students in the fifth, seventh and ninth grades with the highest fitness levels also scored highest on standardized reading and math tests. The students who were the least physically fit had the lowest test scores.

- That same year, a panel of thirteen researchers performed a large-scale review of 850 studies about the effects of exercise on the nation's youth. As reported in *Pediatrics,* the panel concluded that school-age children should participate daily in one hour or more of moderate to vigorous exercise.

- According to a 2009 study published in *Brain Research,* physically fit thirteen- and fourteen-year-olds showed significantly greater brain function than their sedentary peers.

- Physical activity has been shown to boost memory in young women aged eighteen to twenty-five, and it improves frontal lobe function in older adults.

- Exercising protects the short-term memory structures in your brain from high-stress conditions. Stress causes the adrenal glands to produce excessive amounts of the hormone cortisol, which has been found to kill cells in the hippocampus and to impair memory. In fact, people with Alzheimer's disease have higher cortisol levels than normal aging people.

Exercise Alleviates Depression

About 5 percent of children and adolescents experience major depressive disorder. While many people turn to antidepressants, the truth is that exercise may be as effective as prescription medicine in treating depression for some people. One study (2000) compared the benefits of exercise to those of the antidepressant drug Zoloft. After twelve weeks, exercise proved equally as effective as Zoloft in curbing depression. After ten months, exercise had greater beneficial effects than the drug—and without the sometimes debilitating side effects that come with many antidepressant medications.

Exercise Calms Anxiety and Stress

Anxiety disorders affect as many as one in ten young people in America. Millions more of us spend far too much time fretting about the little things. When worry or negative thoughts take over, exercise can provide a welcome distraction. Research shows that high-intensity physical activity can even reduce the incidence of panic attacks. So if you're stressing out about an upcoming test or job interview, or you find yourself constantly replaying an argument you had, lace up your sneakers and go for a run. It can help clear your mind and calm you down.

Exercise Helps Fight Addictions

Have you ever heard the term *a runner's high?* A good long jog releases endorphins in the brain, and these produce a high that's similar to what you might experience from a drug. Studies show that exercise lowers cravings for nicotine and alcohol and reduces withdrawal symptoms.

The Best Exercises for Your Brain

Cardiovascular Exercise

Aerobic exercise is vital to brain health and affects new cell growth. The best aerobic exercise includes a brief warm-up period, twenty to forty-five minutes of sustained moderate to intense physical activity, and a cooldown period. Some evidence suggests that higher-intensity physical activity—even for short periods of time— is also beneficial to the brain. Running, walking fast, swimming, rowing and stair-climbing are just some of the many aerobic exercise options available. And your brain will benefit whether you get your heart pumping outdoors or in the gym. And the best part? You don't need a lot of expensive equipment—just throw on a pair of sneakers and go.

Resistance Training

According to a study (2009) published in the *British Journal of Sports Medicine*, resistance training may prevent cognitive decline in older adults. It also builds strength and tones muscles, and you can use any type of resistance, such as dumbbells, medicine balls, resistance tubing or even your own body weight—for example, push-ups, pull-ups or squats. Some resistance training exercises, like rowing, swimming and stair-climbing, double as aerobic activities, which makes them even better for your brain.

Coordination Exercises

Exercise that requires coordination enhances thinking, cognitive flexibility and processing speed. This means that participating in activities like dancing, tennis and basketball can actually make you smarter! And that's not all. Animal studies have shown that physical exercise that involves the planning and execution of complex movements actually changes the brain's structure.

In 2008 Brazilian researchers compared the brains of competitive judo players and non-judo players. Judo is a form of martial arts that relies on quick reactions and cunning to outmaneuver an opponent. The judo players had significantly higher gray matter tissue density than the non-players, and more gray matter means better brain function.

Combo Exercises

It is a good idea to engage in various types of exercise. Aerobic activity spawns new brain cells, but it is coordination exercises that strengthen the connections between those new cells so your brain can recruit them for other purposes, such as thinking, learning and remembering.

Physical Activities for Your Brain Type	
If you have:	**Try this type of exercise:**
» Prefrontal Cortex Problems: ADHD, short attention span, impulsiveness, poor planning	» *Lots* of high-intensity aerobic activities, meditation
» Cingulate Problems: holding grudges, getting stuck on negative thoughts, OCD	» Intense aerobic activity to boost serotonin
» Deep Limbic Problems: depression, PMS	» Aerobic activity in social/ group settings
» Basal Ganglia Problems: anxiety, panic attacks, constant worry	» Yoga, aerobic activity

SUMMARY

- Physical activity is important for your brain and body.

- Physical activity has been linked to an improvement in academic performance.

- Exercise has been found to be helpful for certain brain struggles (e.g., depression and anxiety), although too much exercise is harmful.

QUESTIONS

1. How would you rate your current physical activity level?

2. Why do you think your physical activity level has decreased over time, if it has?

3. What can you do to increase your level of physical activity?

16 WORKING THE BRAIN TO BE SMARTER

When you think of exercise for the brain, many of us might only think of the physical workouts mentioned in Chapter 15. The truth is, however, that not only do you need to work the muscles in your body, but you also must exercise your brain. Like the saying goes, use it or lose it! In fact, the more you use your brain, the more you can do with it.

Research shows that learning causes new connections to form in the brain. Learning keeps neurons firing and makes it easier for them to fire. Our brain needs stimulation, such as what occurs when reading, to grow and develop in healthy ways and to make it more efficient. When your brain functions better, you are more likely to have an active, learning brain throughout your life.

The scary part is what happens if you do not keep yourself engaged in learning new things: the brain actually begins the process of disconnecting itself! This can negatively affect your academics, your job performance and even your relationships. Doing the same things over and over—even if those things are relatively complex—is bad for the brain. If the most intelligent rocket scientist stopped learning, content with what he or she already knew, even a genius brain like that would begin disconnecting itself.

Now, for the teenager and young adult, not engaging in any new learning is pretty difficult to do. If you are in school, you are forced to learn new information and skills, whether you want to or not. But the attitude you bring to class plays a critical role. Too often, I have seen kids in high school show up for class and not make the slightest attempt to take learning seriously. A vast majority of people look back on their high school years and think, *If only I had taken my education more seriously*...

You might think that such indifference about education changes in college, that students in a university setting would be highly engaged in their studies. Sadly, this is not the case, either. As a professor of education, I'm basically a teacher who teaches future teachers how to teach. Yeah, try and say that ten times fast. In the last two years of our undergraduate program, the classes that my students take are all focused on one thing: how to be effective teachers. These are very concentrated and relevant classes in teaching methods, content pedagogy, classroom management, education psychology, instructional alignment, student teaching and much more. You would think that students at this level would be highly engaged and would take their classes seriously given that they relate specifically to their chosen careers. Nope.

Over the years, I have seen how the brain deeply impacts my students' ability to learn. My students with prefrontal cortex issues are the ones who procrastinate, are easily distracted, and have the hardest time paying attention and focusing on how what they are learning in class is important for when they are teachers themselves. Students with cingulate issues, however, already feel like they know everything it takes to be a teacher. These students are particularly argumentative and stubborn, and they may even debate issues of education in class to show they are right. Students with basal ganglia issues are *terrified* of becoming teachers, because they believe that they are going to ruin all their own students' lives. This anxiety causes them physical pain and makes their student teaching classes absolute nightmares. My students with deep limbic system issues are typically more concerned about their relationships and friendships than they are about learning. And those who are depressed are unmotivated and think, *What's the point? I'm never going to be a teacher, anyway, because I'm such a failure.*

The important lesson to learn here is: while you are young, do your absolute best to engage actively in as much learning as you possibly can! Think about it this way: As you grow up, that three-pound super-computer between your ears goes through a number of critical and developmental periods. It's easier to learn new skills now, when you are younger, because your brain is still developing, maturing and is still determining which parts of itself to prune. That's why it's much easier to pick up new skills, such as a foreign language or a new sport, when you are younger. When you're older, and your brain develop-ment is complete, it becomes much more difficult to master these same tasks.

One way that teenagers and young adults have been hurt is in how they were praised and rewarded when they were children. Think of how many times you have heard an adult tell a child, "Good job on getting that A," "You are so smart," or "So you got a B. Better luck next time." You might think these comments would boost chil-dren's self-esteem or encourage them to do well and keep learning. Unfortunately, this is not the case. When children are praised for the accomplishment, it puts the entire focus on the result, instead of the behavior that enabled the accomplishment—and that behavior is the most important part. What did it take to get an A? Probably lots of hard work and studying, asking questions in class and staying engaged with the material.

Now, imagine if the praise a child received was, "I am really impressed at how much effort you put into your assignments and the hard work you showed in class. Great job!" Notice how the focus is on the desired behavior instead of an end result. As the beneficiary of such praise, the child, even if he or she got a B in the class, will likely remain motivated to work hard and stay engaged.

Researchers at Columbia University compared hundreds of fifth-grade children who were praised for their intelligence with those who were praised for their efforts and hard work. The study (1998) showed that the kids who were praised for being smart became more performance oriented and were less prepared to deal with setbacks. When faced with a challenging task, they were less likely to tackle it with persistence, they exhibited less enjoyment and they did less well on it than the children who were praised for their efforts and hard work. Also, the kids who earned praise for being smart tended to say that intelligence couldn't be improved or developed. Students who were praised for their efforts and hard work believed that they could learn strategies to improve their intelligence and performance.

An important element of intelligence is self-control. Research shows that preschoolers who know how to delay gratification achieve higher academic performance, cope better with stress and frustration, and have better social and cognitive skills as adolescents. Psychology professor and researcher Walter Mischel's famous "marshmallow experiment" illustrates how this works. In the late 1960s Mischel and his colleagues invited dozens of preschoolers into a laboratory room one at a time and had them sit down at a table on which there was a single marshmallow. The researchers told each child that he or she could either eat the marshmallow right away or wait for several minutes and get two marshmallows. Some of the children couldn't wait and ate the marshmallow right away. Other children came up with ways to distract themselves, such as clapping their hands, or to manipulate the environment, such as turning their chair so they faced away from the marshmallow, to keep themselves from eating it.

Mischel followed these children for fourteen years and found that those who were able to delay gratification, to wait for two marshmallows, fared much better in life than those who ate the marshmallow right away. The "waiters" had higher self-esteem, were better at coping with stress and frustration, performed better academically, scored an average of 210 points higher on their SATs and were more socially adept than the "gobblers."

In a follow-up study, Mischel reenacted the experiment, including adults in the process this time. While the youngsters watched, the adults were presented with the single marshmallow and used a variety of tactics to avoid eating it. Then, when it was the children's turn, the ones who had previously eaten the lone marshmallow used the adults' tactics they had just witnessed and successfully managed to wait it out and get the two marshmallows. In later follow-ups, these children performed at levels similar to those who had the natural ability to delay gratification.

What this fascinating study tells us is that children can indeed learn techniques and strategies to improve their intelligence. Teenagers and young adults can reap the benefits of delaying gratification, too. In your younger years, you can—and should—teach yourself self-control and patience. Work to develop your ability to delay gratification and think of the long term. You can take a cue from the children in this study and begin by distracting yourself, to get your mind off of a reward, or by altering your environment, putting the objects of your desire out of sight so that they aren't as tempting for you. Teaching yourself how to be patient can go a long way.

Self-esteem also plays a role in learning. George Valiant, a highly respected professor of psychiatry at Harvard University, has shown that the only thing that correlates to self-esteem as an adult is if a person had performed chores at home as a child. Doing some form of work as a young person can help you develop a sense of competence in adulthood.

Working at a private university, I have seen many cases where children grew up without any real responsibilities. Their parents gave them everything without requiring them to lift a finger and exercise responsibility for something at home. Sometimes, this happens inadvertently, especially in families where the parents might have grown up in poverty. In these cases, the parents try hard to give their children a life that is free from financial stresses, from having to work hard when you're young and worry about where you'll live and what you'll eat. Oftentimes, the parents become upset with their children for not appreciating all the advantages they enjoy. Perhaps you've heard an older person say something like, "When I was young, I had to walk seven miles to school, uphill, in the snow!"

This means that having a part-time job when you are a teenager might be in your best interest. A job can give you a sense of responsibility and maturity, and it can teach you some of the most important skills you will need to possess as an adult. It will increase your self-esteem, giving you a sense of worth and accomplishment. Don't wait to have your first job interview until you are twenty-two years old and already out of college.

Perfect Practice Makes Perfect

Have you heard the expression "Practice makes perfect"? Well, that statement is not entirely accurate. It should be "Perfect practice makes perfect." If I spent hours and hours practicing the wrong way of playing table tennis, it would only make me perfect at playing table tennis the wrong way! If I want to do something efficiently and effectively, I have to learn how to do it right. Then, with time and additional practice doing it correctly, it gets easier and easier, until it isn't much of an effort anymore. It is wired in the brain and becomes a part of who I am.

Long-term potentiation (LTP) is an important factor in why even things that are difficult to master become easier to do over time. It involves invigorating (or potentiating) neurons to do their job over a long period of time. It means that learning happens through the repetition of an act, which causes actual physical changes in neurons and their synapses. LTP causes the nerve endings to get bigger, which creates a significant advantage on three fronts: 1) they are harder to damage; 2) better signals can then pass between each of the cells, which makes it easier for them to communicate; and 3) neurons will then be able to generate their own signals with less input over time. Once LTP has happened, it takes less energy for you to do something well.

For example, when I first sit down at the piano to learn a new song, the process seems daunting and frustrating. Even after years of playing, I still find myself counting keys and looking at the notes to make sure I am playing them right when I learn a new piece. After hours of practice, however, it gets easier. I don't have to think about the notes, placement and rhythm as much. Then, with enough time and practice, I can play the song without having to have any of the notes in front of me. This happens because I steadily stimulate the synapses in the neurons of my brain that control the finger movements required to properly play the piano piece. In essence, I potentiated my neurons in just the right way to master the piano piece. (This also means that my next step is to start working on a new piano piece, if I want to keep sharp.)

Challenge Yourself

Most of us tend to find the things we're good at and stick with them. I know that I am awful at dancing, and therefore, I avoid it at all costs. I can still remember my cousin's *quinceañera,* or fifteenth birthday celebration. To this day, members of my family still fire up that video and make fun of my awkward side shuffle and the look of horror on my face. I am also terrible at playing baseball. When I was an elementary school teacher, I was kicked off our staff team because I struck out, missed fly balls in left field and once accidentally trampled a fifty-five-year-old teacher from a competing school while trying to get to second base. But by avoiding trying to get better at the things I am not good at, I am not doing my brain any favors.

The best way to stay mentally fit is to try the things that don't come naturally to you. If you aren't a dancer, give square dancing or ballroom dancing a whirl. It will fire up areas of your brain you've been neglecting. With practice, you can actually improve your ability to get down with your bad selves. And while you may not win *Dancing with the Stars,* you might feel confident enough to strut your stuff at the prom or the next family wedding.

It is also time to discard the notion that the things we are good at are inherited, innate or even genetic. While genes might play some role in our brain wirings and behavioral tendencies, it is what you do on a day-to-day basis that shapes your brain and develops your talents and abilities.

Fun Ways to Stimulate Your Brain

Studies show that preschoolers who play with good old building blocks are more likely to achieve high math scores in middle school and high school and to improve their language development. Even some traditional board games have been shown to enhance a variety of math skills in preschoolers.

Whenever I tell people to play brain-healthy games for mental exercise, they immediately assume that I'm talking about crossword puzzles or word games like Scrabble or Boggle. While these are great forms of mental exercise that shore up the language centers of the brain, there are many other areas of the brain that need exercise, too. The areas that control memory, concentration, attention, visual-spatial acuity, understanding, planning, anticipation, reaction and coordination can also be strengthened with games. I recommend playing a variety of games that work all the different parts of the brain.

Creative Games That Activate the Right Side of the Brain

For most of us, the left side of the brain, which is dedicated to things like logic and detail, is dominant. Therefore, it's a good idea to stimulate the more creative and artistic right side of the brain with activities like arts and crafts, playing with dolls or puppets and even playing charades.

Word Games That Stimulate the Language Centers

Games like Scrabble and Boggle are great fun and can help teens and young adults learn new words and definitions, enhance language skills, ramp up concentration and improve memory. A bonus is that the scoring in these games also sharpens basic math skills. Crossword puzzles enhance language skills and protect the brain from memory loss. Solving a crossword puzzle as a group also fosters teamwork and patience. You can also try word scrambles or a game as simple as hangman, which challenges the brain in different ways. Even if you feel you are word challenged, you will find that you can get better and sharper with practice.

Coordination Games and Activities

If you want to have some fun while fueling brain cell growth, try juggling. According to a paper published in *Nature* (2004), spending three months learning to juggle sparks growth in certain parts of the brain. German researchers studied twenty-four students over a three-month period: twelve of the students learned a classic three-ball juggling routine, while the other twelve did not learn to juggle. The brains of the jugglers and non-jugglers were scanned before and after the three-month learning period. At the conclusion of the study, the jugglers had more gray matter in the areas of the brain that control memory, language and reading.

Some of the newer video games that involve body- and movement-sensing devices (e.g., Microsoft Kinect) can also be beneficial for brain development, as long as they require you to use hand-eye coordination. (This is a very different kind of video-game activity than playing a first-person shooter, which serves only to activate the dopamine [addictive pleasure] receptors of your brain.)

Games That Involve Music

Music and rhythm are housed in your temporal lobes, which also control memory, reading and language. Teens and young adults can improve these skills by stimulating the temporal lobes with games like karaoke or *Name That Tune*, which enhance your ability to listen and exercise your memory.

Strategy Games That Give Multiple Areas of the Brain a Workout

Playing chess activates several parts of your brain at once. In a study that appeared in *Cognitive Brain Research* (2003), researchers performed MRI scans on chess players, and the scans showed activation of both the left and right hemispheres of the frontal, parietal and occipital lobes. The conclusion is that a single game of chess can stimulate planning, follow-through, attention, impulse control, direction sense and visual-spatial ability.

Memory Games

Simple games, such as the card game Concentration, aren't just for young children. Teens and young adults can also strengthen their memory by playing them. Trivia games also challenge your memory skills and keep you mentally sharp.

Games That Rely on Visual Power

Jigsaw puzzles offer a healthy dose of mental aerobics for teens and young adults. They activate the left side of your brain, the side responsible for noting the details of each puzzle piece, as well as the right side of the brain, which sees the big picture. Jigsaw puzzles improve problem-solving skills, hand-eye coordination and observational abilities, while also strengthening the connections in your memory centers.

10 Ways to Exercise Your Brain

Here are ten different ways that you can exercise your brain to keep it from disconnecting itself:

1. Spend at least fifteen minutes a day—every day—learning something new.

2. If you are in school, make an effort to really learn something in your classes and then contemplate how you can apply it to your life. If you are not in school, take classes on a new and interesting subject at a local Learning Annex or community college.

3. Visit new places, whether it entails taking a vacation or simply a day trip.

4. Join a reading group to help protect your short-term memory.

5. Cross-train in your job. Learn how to do a coworker's job or take on new tasks and responsibilities, rather than doing the same thing over and over.

6. If you're in college, sign up for courses and electives that aren't specifically related to your major. Perhaps a semester of Swedish or ancient Aztec art?

7. Learn to play a musical instrument. If you already play one, try a new one.

8. Challenge yourself to get even better at your favorite activities. Expand your skill set or your knowledge. If you love to scrapbook, for example, try doing more complicated projects. If you play guitar, learn that new hit song you heard on the radio.

9. Surround yourself with smart, interesting people; spending time with people who challenge you intellectually can improve your brain.

10. Shake up your routine. Doing the same thing day after day dulls your brain. Trying a different route to school or work, introducing a new hobby into your repertoire, or finding a new place to hang out or study can stimulate new parts of your brain, make new connections and increase your mental fitness.

SUMMARY

- Mental exercise is very important for your brain.

- It's never too late to learn something new, although it is easier to do this when you are still young.

- You have to keep learning new things to create new connections in your brain.

- When you stop learning, your brain starts to disconnect itself.

- Spend fifteen minutes a day learning something new.

QUESTIONS

1. What are some new things you would like to begin learning?

2. What are ways in which you can dedicate short amounts of time to begin learning these new things?

3. Why is it important to continue learning?

RELAXING
THE BRAIN

Teenagers and young adults today face unprecedented amounts of stress. High-stakes tests, GPAs, schedules, college applications, the economy, romantic relationships and even global safety are just some of the many stresses you may be dealing with on a daily basis.

Whenever you have stress, your body produces a significant biological response. As stress is signaled in the brain, your body begins to release adrenaline and cortisol hormones, which are the primary chemicals of the fight-or-flight response system. When this happens, your heart pounds faster, your breathing gets faster, the blood flows faster through your brain and your mind feels like it is in a more heightened state of alertness. To some degree, this is a good thing, because it means that you'll be ready for just about any situation. Think about how stress affected you during a final exam or a job interview, or on a first date. When we feel this stress, we are motivated to do a good job, study harder or behave the best we can.

Chronic Stress

When the body is stressed too much and for too long, the continuous exposure to adrenaline and cortisol can have devastating effects. Chronic stress has been linked to:

» Alzheimer's disease	» Anxiety
» Blood vessel malfunction	» Cancer
» Decreased brain reserve	» Depression
» Diabetes	» Disruption of sleep patterns
» Heart disease	» High blood pressure (hypertension)
» Increased clotting of the blood	» Inflammation
» Lowered immune system functioning	» Obesity
» Premature aging	» Reduced blood flow
» Susceptibility to disease	

When stress levels rise, we often resort to all sorts of unhealthy ways of coping. We reach for a double cappuccino to help us power through the day. Due to the elevated levels of cortisol in our system, we tend to eat more, and we're more likely to give in to sugar and fat cravings. We skip our daily workouts because we're too busy. We skimp on sleep because we're too wired to doze off at night. We worry about how stressed out we are, which creates even more stress. Caffeine, poor eating habits, lack of exercise and inadequate sleep decrease brain function and lower our ability to deal with stress.

Chronic Stress Makes You Sick

In a 2004 issue of *Psychological Bulletin,* a team of psychologists published findings from a thorough review of nearly three hundred scientific studies on the impact of chronic stress on the immune system. According to their analysis, the studies, which were conducted from 1960 to 2001 and involved 18,941 test subjects, provided incontrovertible evidence that stress causes changes in the immune system. What they found is that short-term stress temporarily boosts immunity, but chronic stress weakens the immune system, making people more vulnerable to common ailments and serious diseases. In particular, the elderly and those who are already suffering from an illness are more susceptible to changes in the immune system due to chronic stress.

How Chronic Stress Plays Tricks with Your Mental Health

Chronic stress drains emotional well-being and is associated with anxiety and depression, which activate the limbic system of the brain. If you experience some form of emotional trauma—say, you are involved in a car accident—your limbic system becomes very active, which can make you more upset and depressed. Some people develop PTSD, which means the stress never goes away.

How Chronic Stress Makes You Fat

One study (2006) found that stress causes people to turn away from healthy, low-fat foods in favor of high-fat junk food. For instance, while under stress, you might ditch the grapes for a bag of M&M'S.

One study (2006) out of Georgia State University showed that when hamsters faced repeated stress over a thirty-three-day period, they overate and gained weight, and in particular, they gained a significant amount of abdominal fat, also known as visceral fat. This type of fat is the worst kind because it surrounds vital organs and is associated with a number of serious diseases, such as cardiovascular disease and diabetes. A study conducted by researchers at

Georgetown University Medical Center in 2008 found that chronic stress packs on even more abdominal fat than you might experience from a high-fat, high-sugar diet alone—and it does it faster.

Several years ago I found myself working for a supervisor who was grossly inept at being a leader and a manager. Within months of assuming his position, he became a running joke around the office. During meetings, he became easily overwhelmed, confused and defensive, and was somewhat abrasive at times. At one point, he expressed concern about an employee because she worked as a clown on the weekends. For whatever reason, he was worried that she might scare little children and drive away business.

Since he was my direct supervisor, I had regular meetings with him to update him on my tasks, projects and responsibilities. For the first several weeks, I tried to remain calm and understanding when he repeatedly said he didn't understand what I was telling him. I repeated my answers to him and then spent hours trying to explain the rationale behind my decisions and plans, but he still wasn't able to process what I was saying. After a while, however, I noticed my stress levels increasing dramatically. There was only so much repeating I could take! Within two months I found myself arguing with him in his office, throwing my hands up in the air, slamming doors and cursing to myself as I walked back to my office.

Other employees in the office also began to exhibit similarly high levels of stress. By the end of three months, several employees had quit, and the morale across the office was abysmal. I didn't have the luxury of quitting, so I endured, determined to fight on. This took a drastic toll on my brain and body. I gained nearly twenty pounds and began having anxiety attacks in the middle of the night. It felt as if my lungs were closing on me as I struggled to breathe. I started to break out in rashes across my face and body, and I became obsessed with the idea that something was terribly wrong with me.

After one particularly frustrating meeting, when I had to explain to my boss how to read a budget report more than ten times, I had stomach pains so intense that I went to the ER to make sure I wasn't dying of some horrific illness. I was sent home with an anti-gas medicine and told to try not to be so stressed out. That's when I realized I was so chronically stressed out and unhappy in my job that it was nearly killing me. After I quit, my brain and body stabilized within a couple of weeks, and I made a promise to myself (and to my wife) never to allow work to stress me out to that degree again.

Who Is Vulnerable to Stress?

Unfortunately, everyone is vulnerable to the effects of chronic stress. When chronic stress hits you or someone in your circle—whether it's a family member, a teacher or a friend—everybody around you or that other person suffers. Perhaps you've heard of the theory of trickle-down economics? This is trickle-down stress. When a parent is stressed out, everybody at home is stressed out.

Young Children and Adolescents

Many adults mistakenly believe that childhood is completely stress free. As you know, growing up is a high-stress endeavor. Young people have to deal with homework, tests and parental pressure to perform well at school. They might have to deal with difficulties making friends, going through romantic breakups or being bullied. Problems at home and changes in family life—such as parents getting divorced or fighting all the time, dealing with a new sibling or a new stepparent, moving—also create stress for teens and young adults. And research shows that living in poverty elevates stress hormone levels in children, which can impair brain development and reduce short-term memory, a process that is essential for language comprehension, reading and problem solving.

College Students and Young Adults

Going to college and starting out on their own is a very exciting time in young people's lives. However, going from a structured home life to the freedom that college life or "real life" offers causes stress. In many cases, young people are experiencing life without their family support system for the first time, and it takes time to create a new support system. Academic, financial and social pressures combine to create intense stress for some. In one study of young adults, researchers found that short, temporary increases in cortisol negatively affected their thinking and memory skills.

6 Brain-Healthy Ways to Combat Stress

The following are helpful ways to fight against the effects that stress can have on your brain and body.

Calming Activity #1: Meditate or Pray on a Regular Basis

Decades of research have shown that meditation and prayer calm stress and enhance brain function. At the Amen Clinics we performed a SPECT study on meditation and found that it significantly increases activity in the prefrontal cortex, which shows that meditation helps tune people in, not out. Here is an example of a simple and effective meditation technique that only takes about 15 minutes but provides great benefits:

- Find a comfortable spot to sit quiet and comfortably and close your eyes.

- Focus on your breathing, and make sure to breathe through your nose. Breathe in slowly and deeply. As you breathe out, say a word silently to yourself.

- Focus on relaxing each of your muscles. You can start at your toes and then move upwards, all the way to the top of your head.

- Continue for about 15 minutes.

Calming Activity #2: Learn to Delegate and to Say No

You don't have to accept every invitation, project or opportunity that comes your way. When someone asks you to do something, your first response might be, "Let me think about it." Then you can take the time to process the request to see if it's something you really want or need to do.

Calming Activity #3: Practice Diaphragmatic Breathing

The simple act of breathing eliminates carbon dioxide from the body. When there's too much carbon dioxide in your system, it can result in feelings of disorientation and panic.

Diaphragmatic breathing calms the basal ganglia, helps your brain run more efficiently, relaxes your muscles, warms your hands and regulates your heartbeat.

This is how you do it: As you inhale, let your belly expand. When you exhale, pull your belly in to push the air out of your lungs.

Diaphragmatic Breathing Exercise #1
1. Lie on your back and place a small book on your belly.
2. When you inhale, make the book go up.
3. When you exhale, make the book go down.

Diaphragmatic Breathing Exercise #2
1. Whenever you feel stressed, anxious or tense, take a deep breath.
2. Hold it for four to five seconds.
3. Slowly exhale, taking about six to eight seconds to exhale completely.
4. Repeat this series about ten times, and odds are that you will start to feel relaxed.

Calming Activity #4: Surround Yourself with the Sweet Smell of Lavender

This popular aroma has been shown to reduce cortisol levels and promote relaxation.

Calming Activity #5: Create a Mental Journal Filled with Positive Memories

In examining the brain scans we take at the Amen Clinics, we have found that when people think about happy memories, it enhances brain function.

Calming Activity #6: Try Self-Hypnosis

Like meditation and prayer, self-hypnosis is a powerful tool for balancing brain function and decreasing stress. When I'm feeling overly stressed, I use the following self-hypnosis exercise. This is one of my favorite stress busters. It usually makes me feel very refreshed and relaxed.

- Focus your eyes on a spot and count to twenty. Allow your eyes to begin to feel heavy as you count and close them when you get to twenty.

- Take three or four very deep breaths, exhaling as slowly as possible with each breath.

- Roll your eyes as far up as you can; then let them return to normal. As they come down, let the muscles in your eyes feel very relaxed.

- Feel how relaxed your eye muscles are and imagine the relaxation spreading like a warm, penetrating oil all over you.

- Imagine yourself walking down a staircase, riding down an escalator or going down in an elevator. As you descend, count backward from ten.

- Imagine that you have arrived at a special haven, a safe spot that's just for you. For me, my special place is a snowy mountain scene where I'm bundled up inside a cozy cabin, in front of a fire. Your special place might be next to a loved one, where you snuggle up, or in a hammock on a warm, breezy beach, where you relax. Experience your special place with all five senses—sight, touch, taste, smell and hearing.

- Focus on three things you're grateful for and stay in your "special place" for about ten to fifteen minutes, and then come back to full consciousness.

SUMMARY

- Chronic stress leads to a whole lot of bad things for your brain and body.
- Meditation/prayer calms and changes the way the brain works.
- It's okay to say no.
- Deep belly breathing really is helpful, if you do it right.
- Certain smells can calm the brain.
- Remember to concentrate on happy memories and thoughts.

QUESTIONS

1. What does chronic stress do to the body?
2. What are the stresses in your life?
3. How can you combat these stresses?

18 THE TWO-WEEK BRAIN SMART PLAN

After everything you've read about the different regions of the brain and the behaviors they affect, and what happens when the brain struggles, it is finally time to put it all together so that you can change your brain and change your life!

For those of you with prefrontal cortex issues, this is probably the first chapter you have actually started reading.

Those of you with overactive cingulates may disagree with much of the information presented thus far, or you might have become obsessive-compulsive about avoiding anything and everything that could potentially cause harm.

Those of you with overactive deep limbic systems might have reacted with hope and a renewed sense of healing and motivation to change when reading the preceding chapters, or you might have shut down even more, feeling that because this is a brain issue, there really is nothing that you are going to be able to do about it.

Finally, if you are struggling with your basal ganglia, you have probably become fairly anxious and fearful, having predicted every worst-case scenario about your brain and your life.

The majority of you, however, have probably responded to this information in the same way that most teenagers and young adults do: you like what you have read, it makes a lot of sense, and you are curious to see what changes you can make (and what you shouldn't do) to better your chances of achieving academic, career and relationship success.

Whatever your motivation has been to read this book—whether you picked it up on your own, it was given to you by another person or you were forced to read it in class—my sincere hope is that you have developed a higher level of respect for your brain and are more committed than ever to taking care of it.

Here are eight key points to remember:

1. Your brain is the most important part of who you are. It is involved in absolutely everything you do. Yes, this includes all your bodily functions, but the kicker is that your brain is also responsible for your thoughts, mood, actions, reactions, interactions, personality, memories, spirituality, joy, feelings, relationships, success, energy, focus, creativity, failures, anxiety, decisions, hurts, dreams and much more.

2. Your brain is ridiculously complex and fragile. It is the most complicated organ in the known universe, is as soft as warm butter and is easily hurt in a vast number of ways.

3. Your brain is not fully developed until you reach about the age of twenty-five. Development starts from the back of the brain, and then works its way to the front. This means that as teenagers, our deep limbic system (emotions) is running rampant and lacks help from a mature prefrontal cortex, which is the last part of our brain to develop fully. For teenagers and young adults, this means that everything you

do, each and every day, has the potential to harm or hurt your brain in a way that can permanently impact your future.

4. Certain parts of the brain are involved in certain behaviors. The prefrontal cortex acts like the supervisor of the brain, the cingulate serves as the gearshift, the deep limbic system acts as the mood stabilizer and the basal ganglia manage movement and anxiety.

5. Normal is definitely not normal. We all have struggles. But this doesn't mean that we have to be held hostage by them. It just means that the brain is not working the way it should be. It is time to get rid of the stigma and judgment associated with mental illness, because the truth is that we all have our own problems.

6. Your brain can change. This is the most exciting part, because it empowers us to know that we are in control. What you do matters, and with the appropriate level of knowledge and insight, your potential is limitless!

7. When we understand how our brain works, we can then look for ways to optimize it to enhance our lives and ensure our success. When we understand how others' personalities and behaviors are shaped by brain function, as well, we can then look for ways to adapt our own behaviors and interactions in order to develop deeper, more meaningful and mutually beneficial relationships.

8. In order to have the best chance for success, you have to make sure that your brain is working right. This means avoiding trauma/injury, drugs, alcohol, malnutrition, chronic stress, infections, environmental toxins, oxygen and

sleep deprivation, smoking, excessive caffeine, too much TV, violent video games, dehydration, negative thinking and a lack of exercise. It also means that you have the responsibility to do everything you can to nurture your brain and take care of it in a way that encourages its health and optimal function, which is what this chapter is all about.

Developing the Two-Week Brain Smart Plan

The following two-week plan is the product of years of experience, research and practice. Over the past thirty years, Dr. Amen has been studying the brain. Along the way he has drawn connections between the brain and human behavior and success. In his work, Dr. Amen has found great success with adults who are already dealing with brain struggles. For the first time, these people have felt like they could truly understand what was happening in their lives. More important, they were given the skills, strategies and insight needed to take control and change their brains and their lives.

When I began working with Dr. Amen more than ten years ago, I quickly realized that we needed to do more. Since my career has been in the field of education, my passion, I already had years of experience working with children with brain struggles. During this time, I often found myself frustrated with the vicious cycle of young people engaging in activities that harmed their brain.

We came to the realization that teenagers and young adults aren't given information about the brain that is practical and relevant to their individual lives and futures. We also had to acknowledge the fact that young people make decisions using a brain that is not even fully developed yet. In other words, there is a reason why teenagers and young adults find it difficult to consider the consequences of their actions, plan ahead and use good judgment before doing something stupid. In the end, we asked ourselves, "How can we expect young people to make smart decisions when their brain is not yet developed *and* we haven't taught them about the brain to begin with?"

When we began talking with teachers, principals and other educators, we became excited by the energy and support we received from them. They, too, had been asking themselves these questions. Soon we had school districts and individual schools across the country asking how they could help us create a program that would help them teach young people about their brain.

Since our first pilot study back in 2005, we have worked hard to fine-tune the process of teaching teenagers and young adults about the brain. We have set up focus groups, conducted research studies and talked with countless students, young adults, parents and professionals to develop a brain-health plan that works, is easy to follow and empowers young people to take their lives into their own hands.

These efforts culminated in the Two-Week Brain Smart Plan, detailed in the pages that follow. If you adhere to the plan's various steps and truly dedicate yourself to understanding, appreciating and nurturing your brain, you will be amazed at how much more focused, hopeful and empowered you will be. Whether you are struggling in your life or just looking to give yourself a tune-up, the strategies here work for everyone. You've already completed the first step, which is to read this book up until this page. Now it's time to take some action!

DAY 1: JOURNALING, GOAL SETTING AND PLANNING, AND BRAIN ROBBERS WORKSHEET

Brain Robbers

The first day of the Two-Week Brain Smart Plan focuses on journaling, goal setting and planning. Just to be clear, this is not the diary type of journaling. It's journaling that allows you to identify, understand and evaluate the things that you do (or don't do) every day that either help or hurt brain function.

There are many ways that you can begin journaling. You can go out and buy a journal and document your activities in it. You can go to www.brain25.com and purchase a specialized journal or download the free journaling PDF files that were created to get you through the next two weeks. Or you can even download the brain25 app on the iOS and Android mobile platforms for an easy and convenient way of staying on track.

Journaling

On Day 1, you will begin documenting what you feed your brain. You will note the foods you ate, the liquids you drank and the approximate number of calories you consumed that day. All this information will be plugged into a Daily Log. Your Daily Log worksheet should look like this:

DAILY LOG

DAY 1

Time	Foods	Calories
Breakfast		
Snack		
Lunch		
Snack		
Dinner		
Snack		

Goal Setting and Planning

On Day 1, you will also complete one of Dr. Amen's popular tools, the One-Page Miracle. You will need to do this only once. It can be found online at www.brain25.com and on the brain25 app, or you can re-create this tool in your own journal. This activity is simple: you merely have to take a moment to think about what you want for your life. For teenagers and young adults, this is particularly important. Think of this activity as a prefrontal cortex booster, since yours is still developing.

MY ONE-PAGE MIRACLE
What Do I Want For My Life?

HEALTH

Emotional health:

Nutrition:

Fitness:

Weight:

RELATIONSHIPS

Family:

Friends:

Boyfriend/Girlfriend:

SCHOOL

Grades:

Long-term goals:

MONEY

Short-term goals:

Long-term goals:

Brain Robbers Worksheet

During the first five days of the Two-Week Brain Smart Plan, you will complete a simple worksheet that will allow you to identify some of the common ways that you rob your brain of the ability to function at its optimal levels. You can find this worksheet online at www.brain25.com and on the brain25 app, or you can re-create this worksheet in your own journal. To complete this worksheet, circle all the things that you did today that qualify as brain robbers. Then, at the bottom of the worksheet, write a brief explanation of how these brain robbers might have harmed you and what you will do to avoid them tomorrow.

BRAIN ROBBERS
(circle the brain robbers below that you struggled with today)

Alcohol use	Drug use	Environmental toxins
Excessive caffeine	Excessive computer use	Excessive cell phone use
Excessive text messaging	Excessive TV	Excessive video games
Fear of failure	Head injury	High stress
Irrational fears	Negative thinking	Not challenging yourself
Poor diet	Poor mental exercise	Poor sleep
Poor water intake	Smoking	Staying in a routine

Reflection: Write a brief explanation of why you think you engaged in some of these activities and how they might have hurt your chances of achieving the goals you set in your One-Page Miracle, and list some of the ways you will try to avoid them tomorrow.

Your Reflection:

_____

_____

_____

DAY 2: EXERCISE PLAN, DAILY LOG (WITH EXERCISE) AND BRAIN ROBBERS WORKSHEET

Today you will develop an exercise plan and routine. You will be encouraged to dedicate thirty minutes of your day to some type of physical activity. You will also identify the brain robbers from your day, and then you will take some time to reflect on how your behaviors and activities helped or hurt you in achieving the goals you set in your One-Page Miracle.

Exercise Plan

By now, you have a clear understanding of how important physical exercise is for your brain and body. Today you will create an exercise plan for the next couple of weeks. Remember that this does not necessarily mean that you have to go to the gym, lift weights or do spin class (although you are more than welcome to). You can choose such activities as taking a brisk walk, dancing, doing aerobics, jogging or playing a team sport (as long as the sport doesn't put you at risk for a brain injury).

You can find the Exercise Plan worksheet at www.brain25.com and on the brain25 app, or you can create one in your journal. The key to the plan is to select specific physical activities, preferably a different one each day, and to specify a time to do each of these activities.

Make your plan, start with at least thirty minutes of exercise today and then stick with it. It's only two weeks. And if you are like most people who start this plan, you will end up exercising much more, even after the two weeks are up.

EXERCISE PLAN

Date	Physical Activity	Scheduled Time
Day 1		
Day 2		
Day 3		
Day 4		
Day 5		
Day 6		
Day 7		
Day 8		
Day 9		
Day 10		
Day 11		
Day 12		

continued on page 248

Date	Physical Activity	Scheduled Time
Day 13		
Day 14		

Daily Log

Plug your nutritional information into the Daily Log the same way you did yesterday, and from today on also log your exercise. Your new Daily Log worksheet should look like this:

DAILY LOG

DAY 2

Time	Foods	Calories
Breakfast		
Snack		
Lunch		
Snack		
Dinner		
Snack		

Exercise Completed Today	Length of Time

Brain Robbers Worksheet

At the end of Day 2 complete a new Brain Robbers worksheet. This will help you see what you did today that wasn't in the best interest of your brain. It also allows you to see how you may or may not have followed your own advice about avoiding certain brain robbers from the day before, and it will provide you another opportunity to do better tomorrow. More important, this worksheet enables you to see how your actions for the day may have helped or hindered your progress toward achieving the goals you set for yourself in your One-Page Miracle.

As you complete the Brain Robbers worksheet at the end of Day 2, ask yourself this, "Did my actions, thoughts and behaviors today get me closer to my goals? Or did I do things that moved me farther away from them?"

DAY 3: DAILY LOG (WITH ANTs) AND BRAIN ROBBERS WORKSHEET

When you get up this morning, look over your One-Page Miracle to remind yourself of your goals. Remember that you can change these goals any time you wish. Then take a brief look at your Exercise Plan to ensure that you keep on track today.

Daily Log

At the end of the day, complete your Daily Log, which already includes nutrition and exercise. Today you will begin logging a few of the ANTs that you may have had, as well. Refer to Chapter 13 if you need a refresher on ANTs. Also, you can visit www.brain25.com for additional resources.

Your new Daily Log worksheet should include an ANTs section that looks like this:

DAILY LOG

DAY 3

Time	Foods	Calories
Breakfast		
Snack		
Lunch		
Snack		
Dinner		
Snack		

Exercise Completed Today	Length of Time

ANT (write down the negative thought)	ANT Species (identify the species)	ANTeater (talk back to the thought)

Brain Robbers Worksheet

At the end of your day, complete a new Brain Robbers worksheet, identifying any of the things that you did that day that were brain robbers. Reflect on how these kept you from making progress toward your goals, and focus on how to prevent them from happening again tomorrow.

DAY 4: DAILY LOG (WITH GRATITUDE) AND BRAIN ROBBERS WORKSHEET

Start today like you did yesterday. When you wake up, take a look at your One-Page Miracle and your Exercise Plan to make sure you reach your goals for the day.

Daily Log

As you have probably noticed, we are progressively adding information to your Daily Log each day. You have probably become much more aware of your brain after logging your nutrition, physical exercise and negative thoughts. Today you will add to your log one thing that happened that was positive, that you are grateful for. As you might recall from Chapter 13, this is an important task to do because of the power it can have over your brain function, moods and emotions.

Your Daily Log worksheet for today should include a Gratitude section that looks like this:

DAILY LOG

DAY 4

Time	Foods	Calories
Breakfast		
Snack		
Lunch		
Snack		
Dinner		
Snack		

Exercise Completed Today	Length of Time

ANT (write down the negative thought)	ANT Species (identify the species)	ANTeater (talk back to the thought)

Brain Robbers Worksheet

At the end of your day, complete the Brain Robbers worksheet to highlight any key areas of growth you still need to work on. If you cannot find anything on the list that robbed your brain of its potential today, then you can use the Reflection section of the worksheet to describe what you are enjoying and not enjoying about this two-week plan.

DAY 5: CALMING PLAN, DAILY LOG (WITH CALMING) AND BRAIN ROBBERS WORKSHEET

Today you will start your day by creating a Calming Plan for yourself. This worksheet will allow you to plan a ten-minute period of time in your day (starting today and continuing until Day 14) to relax, take in the sights and sounds of the present, breathe deeply and let go of everything else. You should spend this time in a place free from distractions. No music, no friends, no bothers. Since yoga, meditation and prayer promote relaxation and a sense of well-being, they can count as your calming activity.

CALMING PLAN

Date	Calming Activity and Possible Location	Scheduled Time
Day 1		
Day 2		
Day 3		
Day 4		
Day 5		
Day 6		
Day 7		
Day 8		
Day 9		
Day 10		
Day 11		
Day 12		
Day 13		
Day 14		

Daily Log

Your Daily Log should already include details about your nutrition, exercise, ANTs and gratitude. Today you will add to your log the activity you engaged in to calm your brain. Your Daily Log worksheet for today should include a Calming section that looks like this:

DAILY LOG

DAY 5

Time	Foods	Calories
Breakfast		
Snack		
Lunch		
Snack		
Dinner		
Snack		

Exercise Completed Today	Length of Time

ANT (write down the negative thought)	ANT Species (identify the species)	ANTeater (talk back to the thought)

continued on page 256

ANT (write down the negative thought)	ANT Species (identify the species)	ANTeater (talk back to the thought)

Gratitude (write down one thing that you are grateful for today)

Calming Activity for Today	Length of Time

Brain Robbers Worksheet

Today is the last day that you will complete the Brain Robbers worksheet. By now, you should already have a clear idea of the things you do each day that threaten your brain's potential. From now on, you are on your own to fight against these brain robbers.

DAY 6: DAILY LOG (WITH NEW LEARNING)

Begin your day by taking a brief look at your One-Page Miracle, your Exercise Plan and your Calming Plan to make sure that you stay on track for the day. Then think of something new that you would like to learn. This could be learning to play a new instrument, speak a new language, crochet or anything else. Starting today, you will dedicate fifteen minutes to your new project, and you'll document it in your Daily Log at the end of the day. Today you will add the last section to your Daily Log. After today we will not be adding any more to it. Your Daily Log worksheet for today should include a New Learning section that looks like this:

DAILY LOG

DAYS 6–14

Time	Foods	Calories
Breakfast		
Snack		
Lunch		
Snack		
Dinner		
Snack		

Exercise Completed Today	Length of Time

ANT (write down the negative thought)	ANT Species (identify the species)	ANTeater (talk back to the thought)

continued on page 258

ANT (write down the negative thought)	ANT Species (identify the species)	ANTeater (talk back to the thought)

Gratitude
(write down one thing that you are grateful for today)

Calming Activity for Today	Length of Time

New Learning Activity for Today	Length of Time

Notes

DAY 7: DAILY LOG
AND WEEK 1 REFLECTION

Congratulations! Today marks the end of the first week of working toward a better brain and a better life! Today you will complete your Daily Log as usual. At the end of the day, however, spend fifteen minutes reflecting and writing a summary of what this week has looked like for you. Be sure to jot down what you have loved, what you have hated and what you have learned. This will be important to note at the end of week 2.

DAY 8: DAILY LOG
AND HEALTHY EATING

Spend time today completing your Daily Log and then making a list of the meals that you plan to eat over the next six days. Refer to Chapter 14 for more details and assistance.

DAY 9: DAILY LOG
AND OTHER PEOPLE'S BRAINS

Make sure you complete your Daily Log today and then spend a few minutes concentrating on the three most important people in your life. As you think of these three people, reflect on their traits, personalities, emotions and abilities from a brain science perspective. Try to determine which of their brain systems might be doing well and which ones might be overactive or underactive. Then write down some ways that you can change your own behavior in order to build a deeper and more meaningful relationship with them.

DAY 10: DAILY LOG AND POWER OFF

Today may be the most difficult for some of you. You will take *one day* to unplug your brain and see how it feels. Do not watch TV, play any video games, use your computer/laptop/tablet or touch your phone, if at all possible. Spend the day interacting with people and taking in the sights and sounds of the world around you. Chances are, you will end up figuring out something more creative to do with your brain. At the end of the day, be sure to complete your Daily Log.

DAY 11: DAILY LOG AND SLEEP

Today you will complete the Sleep Journal worksheet to give you a better idea of how you are sleeping and the potential impact of your sleep on your brain. At the end of the day, complete your Daily Log. Make sure that you are still reviewing your One-Page Miracle, recording your nutrition and ANTs, and sticking to your exercise, gratitude, calming and new learning plans, as well.

MY SLEEP JOURNAL

Last night my bedtime ritual included (list things like a warm bath, meditation, reading, etc.):

Last night I went to bed at: _____ p.m./a.m.

Last night I fell asleep in: _____ minutes.

Last night I woke up: _____ times.

During those times, I was awake for: _____ minutes.

Last night I got out of bed: _____ times.

Things that disturbed my sleep included (list any physical, mental, emotional or environmental factors that affected your sleep):

I slept for a total of: _____ minutes.

I got out of bed today at: _____ a.m./p.m.

Upon waking, I felt: __refreshed __groggy __exhausted.

DAY 12: DAILY LOG AND MUSIC

Today try to incorporate as much new music as you can into your leisure time. In fact, you should do your best to try Mozart or some other classical music, but only if that music is new to you. As crazy as it sounds, I have found it fun to listen to classical music as I drive, making up my own words to the melody. It often results in some odd looks from others, but it can keep you on your toes! Make sure you complete your Daily Log at the end of the day.

DAY 13: DAILY LOG, LAUGH AND LOVE

Today make a concentrated effort to be with the people you love and care about the most. Be sure to hug them, smile and express your love often. And try to laugh a lot! You might even consider going to a comedy show or visiting some funny websites with friends and loved ones. Try to engage in activities that make you laugh. Love and laughter are important, because they produce a specific set of brain chemicals that nourish the brain and immune system.

DAY 14: DAILY LOG AND FINAL REFLECTION

You've made it to the very end! Today it is important to spend time reflecting on any changes that you feel have occurred. What is different about your life, relationships, friendships, mood, focus, concentration, happiness, academics and success? What is better? What has been the most difficult aspect of this journey?

Brain Systems Quiz (One Last Time)

You completed the Brain Systems Quiz back in Chapter 1 and then again after you learned all about the brain. Now that you are at the end of your two-week journey toward a better brain, it's time to complete this quiz one last time.

BRAIN SYSTEMS QUIZ

Please rate yourself on each of the behaviors listed below using the scale provided. If possible, have another person who knows you well (e.g., a parent, a significant other, a close friend) rate you, as well, in order to construct the most complete picture.

0	1	2	3	4
Never	Rarely	Occasionally	Frequently	Very Frequently

_____ 1. Failure to pay close attention to details; tendency to make careless mistakes

_____ 2. Trouble sustaining attention in routine situations (e.g., homework, chores)

_____ 3. Trouble listening

_____ 4. Failure to finish things; tendency to procrastinate

_____ 5. Poor time organization

_____ 6. Tendency to lose things

_____ 7. Tendency to be easily distracted

_____ 8. Poor planning skills and a lack of clear goals or forward thinking

_____ 9. Difficulty expressing empathy for others

_____ 10. Impulsiveness (saying or doing things without thinking first)

_____ 11. Excessive or senseless worrying

_____ 12. Upset when things do not go your way

_____ 13. Upset when things are out of place

_____ 14. Tendency to be oppositional or argumentative

_____ 15. Tendency to have repetitive negative thoughts

_____ 16. Tendency toward compulsive behaviors

_____ 17. Intense dislike for change

_____ 18. Tendency to hold on to grudges

_____ 19. Upset when things are not done a certain way

_____ 20. Tendency to say no without first thinking about a question

_____ 21. Frequent feelings of sadness or moodiness

_____ 22. Negativity

_____ 23. Decreased interest in things that are usually fun or pleasurable

_____ 24. Feelings of hopelessness about the future

_____ 25. Feelings of worthlessness, helplessness or powerlessness

_____ 26. Feelings of dissatisfaction or boredom

_____ 27. Crying spells

_____ 28. Sleep changes (too much or too little)

_____ 29. Appetite changes (too much or too little)

_____ 30. Chronic low self-esteem

_____ 31. Frequent feelings of nervousness or anxiety

_____ 32. Symptoms of heightened muscle tension

_____ 33. Tendency to predict the worst

_____ 34. Conflict avoidance

_____ 35. Excessive fear of being judged or scrutinized by others

_____ 36. Excessive motivation (e.g., can't stop working)

_____ 37. Tendency to freeze in anxiety-provoking situations

_____ 38. Shyness or timidity

_____ 39. Sensitivity to criticism

_____ 40. Fingernail biting or skin picking

ANSWER KEY

Questions 1–10 = Prefrontal cortex symptoms

Questions 11–20 = Cingulate symptoms

Questions 21–30 = Deep limbic system symptoms

Questions 31–40 = Basal ganglia symptoms

If you answered two to three questions related to a particular brain system with a 3 or 4, struggles in that part of the brain may be possible. If you answered four to five questions with a 3 or 4, problems in that brain system are probable. If you answered six or more questions with a 3 or 4, problems in that brain system are highly probable.

After you have reflected on this journey, write out your thoughts on paper or on the computer. The process of articulating your experiences can help you process your thoughts, feelings and emotions. If you are willing, upload your document chronicling your experiences to www.brain25.com and share them with others! We would love to hear your comments, feedback and suggestions for making things better for the future.

Conclusion

I'd like to leave you with a story from one of my students. When I first met this student, I could tell that there was something uniquely special about him. And even though he was incredibly gifted, intelligent and passionate about what he wanted to accomplish in his life, I could tell that he was struggling with something deep down inside.

In our very first advising meeting, I began the process of talking about the brain and asking questions about any potential struggles that he might be facing. Within a couple of weeks, he broke down and let me know what had been going on in his life. My heart broke to hear of how much pain, struggle and hurt he had been carrying with him for so long. He felt alone, isolated and always in fear of losing control and being judged by others. To himself, he was a loser with OCD. To me, he was an amazing student with a brain that was struggling.

Luckily, he was determined to make a change. He learned about his brain and started doing things to take better care of it. Within just a few months, he was already feeling more joy and hope for a brighter future. I asked him to share a bit of his experience with me to include in this book. Here is what he wrote:

I remember the first time I had an OCD thought. I was five or six years old and was finishing up playing with my action figures when it occurred to me that if I did not put Spider-Man back in the top drawer of my toy box, I would be attacked by spiders. I remember the anxiety this thought brought about, and the horrible realization that the only way to make it go away was to give in to the urge to ritualize.

As I grew older, the object of obsession changed many times. But the emotional thought process didn't. By high school, I was so terrified of becoming mentally disabled (as if mental disability was an infectious disease) that I was washing my hands over one hundred times per day. My helplessness against the obsessions and compulsions left me feeling isolated and depressed. I was aware that something had run amok in my mind, but I had no idea what. In my mind, I was the only one who had ever suffered from this destructive thought pattern, and there was no chance or hope of improving. As you can imagine, this drew me even further into the grips of depression.

Eventually I would seek help, because I finally accepted something was very wrong with me. I was relieved to learn that what I was afflicted with was a common problem: obsessive-compulsive disorder mixed with depression. I spent a lot of time learning about the brain from Dr. Payne, and I learned that my overactive cingulate was the root of my obsessive thoughts, and an overactive deep limbic system was what made me feel so helpless at times. I remember feeling like the veil had been lifted, and underneath my problems weren't as big as they had originally seemed. Knowing that awry brain structures were the cause of my symptoms brought me a high level of inner peace and hope.

Now I am starting my second year of college, and I feel more equipped than ever to stand up to my obsessions and compulsions. Knowing how my brain operates has given me more insight into my struggles. I know the value of talking down intrusive thoughts and battling strong emotions with logic. I know that some days life is going to be rough, but now that I understand how my brain operates, I have more hope than ever before.

This story is powerful, and it is personal to me because it highlights the turnaround that can happen once people place their brains at the center of their lives. When I met this student, he was battling intense OCD rituals and had been spiraling deeper into depression. Within three months, his entire life changed. You can imagine how excited I will be to see how he has progressed in three years!

My sincere hope is that if you are a teenager or a young adult reading this book, you have become empowered by what you've learned about the brain and how you can change the way it works in your life. At no other point in your life will there be such a fantastic opportunity to rewire your brain for greater success and a better life. If you are able to tap into your mental potential and utilize practical strategies and skills to make your brain great, you will find it easier to solve your problems, create more opportunities and stack the deck in your favor for a lifetime of success.

When Dr. Amen wrote *Change Your Brain, Change Your Life* fifteen years ago, he had no idea that it would help more than 850,000 readers change their lives. This book is an extension of that vision, and it's geared to the population that needs it most.

I am proof of the miraculous changes that come about when someone dedicates the time and energy needed to nurture and enhance the brain. When I have shared moments of my past with others, the response is always the same: "You shouldn't be here. With everything you've been through, statistics say that you should be dead or in some serious trouble." Many times I have been brushed off as an anomaly, an outlier or an oddity who does not represent what typically happens to people in my situation.

I find this ludicrous. The way I see it, I am nobody special. I am no better than anyone else, and I don't have some magical sense of resilience floating around in my DNA. Instead, I just know that I get one life to live on this earth. Even when I was a small child, I always knew that I wanted more. I wanted to be happy, I wanted to take care of myself and I wanted to help others. In the past decade I have been humbled, blessed and grateful to have been given the opportunity to use my experiences, stories and struggles as a platform to give others hope.

So, at the very end, it all rests with you. Now that you have acquired an incredible amount of information about the brain and have learned some of the amazing skills and strategies that have been shown to change your brain and change your life, the question is, What are you going to do about it?

The people who use their brain struggles as a crutch or an excuse not to try harder rarely end up successful. They just end up bitter and continue to blame others for their mistakes. Those who dwell in their past and on negative thoughts, and who allow themselves to be consumed by sadness, end up fulfilling their depression-laden prophecies. Those who are controlled by fear and anxiety never advance, because they don't take the risks that are needed to grow.

In the end, you have the brain that you have. Nothing will change that. The good news is that you *do* have the power to do something with it. Research has shown time and time again that our brains can change. Now you know how to do it. When you accept this reality and dedicate the time and energy to changing your brain and changing your life, amazing things await!

Acknowledgments

First and foremost, I give thanks to God for His steadfast and unconditional love, and for blessing me with the incredible people who have worked tirelessly to make me who I am.

To my wife, Bre: There are no words to describe the gratitude I feel to you. I am who I am and where I am in life because of you. You saw light and hope in me when I didn't want to, and you pushed me to continue on when I was tired, weak and ready to resign.

To my children: Eli, you are my reason for continuing to work as hard as I do. My only hope is to provide you with a foundation that is built on unconditional love and joy so that you can take the reins of your own life and create your own destiny. Emmy, you have taken my heart and made it yours. Your amazing amount of strength, resilience, perseverance and endurance is nothing short of miraculous. You have taught me so much in just a few years, and I am eager to see how you will continue to exceed expectations and be a light for others.

To my family: You lifted me out of my darkest hours and kept me from enduring pain. Tiffany, we've been through a lot, but we've never given up. Lizzy, you've kept me laughing and have always stopped me from taking myself too seriously. Liz, Margie and Lupe, each of you filled the critical parental role at the different points in my life when I needed guidance most. Mom, I know you've always had my best interests in mind and that you've always had the best of intentions. Dad, I thank you for teaching me about perseverance and hard work. I miss you.

To Daniel: Thanks for looking at my brain scan years ago and not judging me or demanding that I stay away from your daughter. Your knowledge, insight and experience changed my brain and my life, and I can only hope to carry on the legacy of your work even further.

Rosie, you saw something in me that was greater than a job in retail. Linda, Ghislaine, Kris and Renee, you were the best supervisors and helped me as I battled my own emotional chaos. Gretchen, you had faith in my ability to teach, even as I was still figuring out how to. Doug, you set the bar high for me and never faltered in your expectations. Jeff, you saw me through many phases of my career and introduced me to the world of higher education. Nick, Rudy, Juanito, Steven and Chavez, thanks for listening.

I would like to thank the many people at Corban University for their continued support, understanding and encouragement. Claudia and Janine, thanks for entrusting me to lead the undergrads and for keeping me grounded and sane. Matt and Kristin, thanks for being incredible and authentic leaders and mentors. Cara and Chelsea, thanks for enduring my nonsense and for being my support and follow-through. Travis, Amy, Caleb, Kent, Drew and David, thanks for enabling me to see my calling here. Vinny, thanks for the edits, feedback and conversations, and for helping to keep me out of my depth.

Finally, I would like to thank the people who helped to make this book happen. First and foremost, a massive thanks to Daniel for initiating this entire process, and for standing by me and supporting me in every way. A big thanks to Celeste Fine for being an amazing agent and working hard to get this into the hands of as many people as possible. Thanks to Liz Neoporent for helping to create a fantastic book proposal. Another big thanks to Sarah Pelz at Harlequin for seeing the potential in this book and for walking me through this very complex and stressful process. Thank you, Liz Stein, for cleaning up my ramblings, getting everything into shape, and allowing me to concentrate more on the content than on minor details.

References

Abdel-Dayem, H. M., et al. (1998). SPECT brain perfusion abnormalities in mild or moderate traumatic brain injury. *Clinical nuclear medicine, 23*(5), 309–17.

Abu-Judeh, H., et al. (1999). SPECT brain perfusion imaging in mild traumatic brain injury without loss of conscienceness and normal computed tomography. *Nuclear medicine communications, 20*(6), 305–10.

Accordino, M., & Hart, C. L. (2006). Neuropsychological deficits in long-term frequent cannabis users. *Neurology, 67*(10), 1902. doi: 10.1212/01.wnl.0000249081.67635.ef

Amen, D. G. (1994). New directions in the theory, diagnosis, and treatment of mental disorders: The use of SPECT imaging in everyday clinical practice. In L. F. Koziol & C. E. Stout (Eds.), *The Neuropsychology of Mental Disorders* (pp. 286–311). Springfield, IL: Charles C. Thomas.

Amen, D. G. (1997). Three years on clomipramine: Before and after brain SPECT study. *Annals of Clinical Psychiatry, 9*(2), 113–16.

Amen, D. G. (1998). Brain SPECT imaging in psychiatry. *Primary Psychiatry, 5*(8), 83–90.

Amen, D. G. (1999). Brain SPECT imaging and ADD. In J. A. Incorvaia, et al. (Eds.), *Understanding, Diagnosing, and Treating AD/HD in Children and Adolescents: An Integrative Approach* (pp. 183–96). Northvale, NJ: Jason Aronson.

Amen, D. G. (2000). *Change Your Brain, Change Your Life*. New York: Three Rivers Press.

Amen, D. G. (2001). *Healing ADD*. New York: Putnam.

Amen, D. G. (2002). *Healing the Hardware of the Soul*. New York: Free Press.

Amen, D. G. (2005). *Making a Good Brain Great*. New York: Harmony Books.

Amen, D. G., & Routh, L. C. (2003). *Healing Anxiety and Depression*. New York: Putnam.

Amen, D. G., Taylor, D. V., Ojala, K., Kaur, J., & Willeumier, K. (2013). Effects of brain-directed nutrients on cerebral blood flow and neuropsychological testing: A randomized, double-blind, placebo-controlled, crossover trial. *Advances in Mind-Body Medicine, 27*(2), 24–33.

Amen, D. G., Wu, J. C., & Carmichael, B. L. (2003). The clinical use of brain SPECT imaging in neuropsychiatry. *Alasbimn Journal, 5*(19). http://www.2.alasbimnjournal.cl/alasbimn/CDA/sec_b/0,1206,SCID%253D3212,00.html

American Psychological Association. (2000). *Diagnostic and Statistical Manual of Mental Disorders, DSM-IV-TR* (4th ed.). New York: American Psychiatric Association.

Arnone, D., Barrick, T. R., Chengappa, S., Mackay, C. E., Clark, C. A., & Abou-Saleh, M. T. (2008). Corpus callosum damage in heavy marijuana use: Preliminary evidence from diffusion tensor tractography and tract-based spatial statistics. *Neuroimage, 41*(3), 1067–74. doi: 10.1016/j.neuroimage.2008.02.064. Epub 2008 Mar 14.

Autti, T., Sipila, L., Autti, H., & Salonen, O. (1997). Brain lesions in players of contact sports. *Lancet, 349,* 1144.

Atherton, M., Zhuang, J., Bart, W., Hu, X., & He, S. (2003). A functional MRI study of high-level cognition. I. The game of chess. *Cognitive Brain Research, 16,* 26–31.

Babbs, C. (2001). Biomechanics of heading a soccer ball: Implications for player safety. *Scientific World Journal, 1,* 281–322.

Barnes, B., et al. (1998). Concussion history in elite male and female soccer players. *American Journal of Sports Medicine, 26*(3), 433–38.

Bava, S., Jacobus, J., Thayer, R., & Tapert, S. (2012). Longitudinal changes in white matter integrity among adolescent substance users. *Alcoholism: Clinical & Experimental Research, 37*(1), 181–89.

Berk, L. S., et al. (1989). Neuroendocrine and stress hormone changes during mirthful laughter. *American Journal of Medical Sciences, 298*(6), 390–96.

Bogg, T., & Roberts, B. W. (2004). Conscientiousness and health-related behaviors: A meta-analysis of the leading behavioral contributors to mortality. *Psychological Bulletin, 130*(6), 887–919.

Bolla, K. I., Brown, K., Eldreth, D., Tate, K., & Cadet, J. L. (2002). Dose-related neurocognitive effects of marijuana use. *Neurology, 59,* 1337–43.

Boll, T., et al. (1983). Mild head injury. *Psychiatric Development, 1*(3), 263–75.

Bridgett, D. J., & Cuevas, J. (2000). Effects of listening to Mozart and Bach on the performance of a mathematical test. *Perceptual and Motor Skills, 90,* 1171–75.

Brody, A. L., et al. (2001). Regional brain metabolic changes in patients with major depression treated with either paroxetine or interpersonal therapy: Preliminary findings. *Archives of General Psychiatry, 58*(7), 631–40.

Camargo, E. E. (2001). Brain SPECT in neurology and psychiatry. *Journal of Nuclear Medicine, 42*(4), 611–23.

Carey, P. D., et al. (2004). Single photon emission computed tomography (SPECT) of anxiety disorders before and after treatment with citalopram. *BMC Psychiatry, 4*(1), 30.

Catafau, A. M., et al. (1996). Regional cerebral blood flow pattern in normal young aged volunteers: A 99mTc-HMPAO SPET study. *European Journal of Nuclear Medicine, 23*(10), 1329–37.

Catafau, A. M., et al. (2001). SPECT mapping of cerebral activity changes induced by repetitive transcranial magnetic stimulation in depressed patients. A pilot study. *Psychiatry Research, 106*(3), 151–60.

Chang, L., et al. (2000). Effect of ecstasy [3,4-methylenedioxymethamphetamine (MDMA)] on cerebral blood flow: A Co-registered SPECT and MRI study. *Psychiatry Research, 98*(1), 15–28.

Chein, J., Albert, D., O'Brien, L., Uckert, K., & Steinberg, L. (2010). Peers increase adolescent risk taking by enhancing activity in the brain's reward circuitry. *Developmental Science*, pp. F1–F10.

Chiron, C., et al. (1992). Changes in regional cerebral blood flow during brain maturation in children and adolescents. *Journal of Nuclear Medicine, 33*(5), 696–703.

Christakis, D. A., et al. (2004). Early television exposure and subsequent attentional problems in children. *Pediatrics, 113*, 708-13.

Colver, A., & Longwell, S. (2013). New understanding of adolescent brain development: Relevance to transitional healthcare for young people with long term conditions. *Archives of Disease in Childhood, 98*, 902–07.

Cotman, C. W., & Berchtold, N. C. (2002). Exercise: A behavioral intervention to enhance brain health and plasticity. *Trends in Neuroscience, 25*, 295–301.

Crippa, J. A., et al. (2009). Cannabis and anxiety: A critical review of the evidence. *Human Psychopharmacology, 24*(7), 515–23.

Datar, A., & Nicosia, N. (2012). Junk foods in schools and childhood obesity. *J Policy Anal Manage, 31*(2), 312–37.

Davidson, J. R., et al. (2004). Fluoxetine, comprehensive cognitive behavioral therapy, and placebo in generalized social phobia. *Archives of General Psychiatry, 61*(10), 1005–13.

De Mendelssohn, A., Kasper, S., & Tauscher, J. (2004). Neuroimaging in substance abuse disorders. *Nervenarzt, 75*(7), 651–62.

Demir, B., et al. (2002, June). Regional cerebral blood flow and neuropsychological functioning in early and late onset alcoholism. *Psychiatry Research*, pp. 115–25.

Di Castelnuovo, A., et al. (2002). Meta-analysis of wine and beer consumption in relation to vascular risk. *Circulation, 105*, 2836–44.

Diamond, M., Cusack, S., & Thompson, W. (2003). *Mental Fitness for Life: A 7 Step Guide to Healthy Aging.* Toronto: Key Porter Books.

Domino, E. F., et al. (2000). Nicotine effects on regional cerebral blood flow in awake, resting tobacco smokers. *Synapse, 38*(3), 313–21.

Dormehl, I. C., et al. (1999). SPECT monitoring of improved cerebral blood flow during long-term treatment of elderly patients with nootropic drugs. *Clinical Nuclear Medicine, 24*(1), 19–34.

Draganski, B., et al. (2004). Neuroplasticity: Changes in grey matter induced by training. *Nature, 427*(6972), 311–12.

Drummond, S. P., et al. (1999). Sleep deprivation-induced reduction in cortical functional response to serial subtraction. *Neuroreport, 10*(18), 3745–48.

Dupont, R. M., et al. (1996). Single photon emission computed tomography with iodoamphetamine-123 and neuropsychological studies in long-term abstinent alcoholics. *Psychiatry Research, 67*(2), 99–111.

Ernst, T., et al. (2000). Cerebral perfusion abnormalities in abstinent cocaine abusers: A perfusion MRI and SPECT study. *Psychiatry Research, 99*(2), 63–74.

Field, A. S., et al. (2003). Dietary caffeine consumption and withdrawal: Confounding variables in quantitative cerebral perfusion studies? *Radiology, 227*(1), 129–35.

Fishbein, M., Gov, S., Assaf, F., Gafni, M., Keren, O., & Sarne, Y. (2012). Long-term behavioral and biochemical effects of an ultra-low dose of tetrahydrocannabinol (THC): Neuroprotection and ERK signaling. *Experimental Brain Research, 221*(4), 437.

Foster, M., Solomon, M., Huhman, K., & Bartness, T. (2006). Social defeat increases food intake, body mass, and adiposity in Syrian hamsters. *American Journal of Physiology – Regulatory, Integrative and Comparative Physiology, 290.*

Furmark, T., et al. (2002). Common changes in cerebral blood flow in patients with social phobia treated with citalopram or cognitive-behavior therapy. *Archives of General Psychiatry, 59*(5), 425–33.

Ghatan, P. H., et al. (1998). Cerebral effects of nicotine during cognition in smokers and non-smokers. *Psychopharmacology 136*(2), 179–89.

Goldapple, K., et al. (2004). Modulation of cortical-limbic pathways in major depression: Treatment-specific effects of cognitive behavior therapy. *Archives of General Psychiatry, 61*(1), 34–41.

Golden, Z. L., et al. (2002). Improvement in cerebral metabolism in chronic brain injury after hyperbaric oxygen therapy. *International Journal of Neuroscience, 112*(2), 119–31.

Goto, R., et al. (1998). A comparison of Tc-99m HMPAO brain SPECT images of young and aged normal individuals. *Annals of Nuclear Medicine, 12*(6), 333–39.

Gudlowski, Y., & Lautenschlager, M. (2008). Impact of cannabis consumption on brain development and the risk of developing psychotic disorders. *Gesundheitswesen, 70*(11), 653–57. doi: 10.1055/s-0028-1100396. Epub 2008 Nov 27. Review. German.

Guszkowska, M. (2004). The effects of exercise on anxiety, depression and mood states. *Psychiatria Polska, 38*(4), 611–20.

Hancox, R. J., Milne, B. J., & Poulton, R. (2004). Association between child and adolescent television viewing and adult health: A longitudinal birth cohort study. *Lancet, 364*(9430), 257–62.

Harris, G. J., et al. (1999). Hypoperfusion of the cerebellum and aging effects on cerebral cortex blood flow in abstinent alcoholics: A SPECT study. *Alcoholism—Clinical and Experimental Research, 23*(7), 1219–27.

Hoecker, C., et al. (2002). Caffeine impairs cerebral and intestinal blood flow velocity in preterm infants. *Pediatrics, 109*(5), 784–87.

Holman, B. L., et al. (1991). Brain perfusion is abnormal in cocaine-dependent polydrug users: A study using technetium-99m-HMPAO and SPECT. *Journal of Nuclear Medicine, 32*(6), 1206–10.

Holthoff, V. A., et al. (1999). Changes in regional cerebral perfusion in depression. SPECT monitoring of response to treatment. *Nervenarzt, 70*(7), 620–26.

Ingram, D. K., et al. (2004). Development of calorie restriction mimetics as a prolongevity strategy. *Annals of the New York Academy of Science, 1019*, 412–23.

Iyo, M., et al. (1997). Abnormal cerebral perfusion in chronic methamphetamine abusers: A study using 99MTc-HMPAO and SPECT. *Progress in Neuro-Psychopharmacology and Biological Psychiatry, 21*(5), 789–96.

Jacini, W., Cannonieri, G., Fernandes, P., Bonilha, L., Cendes, F., & Li, L. (2008). Can exercise shape your brain? Cortical differences associated with judo practice. *Journal of Science and Medicine in Sport.* doi: 10.1016/j.jsams.2008.11.004

Johnson, D., et al. (1999). Cerebral blood flow and personality: A positron emission tomography study. *American Journal of Psychiatry, 156*(2), 252–57.

Johnstone, B., Childers, M. K., & Hoerner, J. (1998). The effects of normal ageing on neuropsychological functioning following traumatic brain injury. *Brain Injury, 12*(7), 569–76.

Jones, R. (2004). Juggling boosts the brain. *Nature.* doi:10.1038/nrn1357

Kao, C. H., Wang, S. J., & Yeh, S. H. (1994). Presentation of regional cerebral blood flow in amphetamine abusers by 99Tcm-HMPAO brain SPECT. *Nuclear Medicine Communications, 15*(2), 94–98.

Keightley, M., et al. (2003). Personality influences limbic-cortical interactions during sad mood induction. *Neuroimage, 20*(4), 2031–39.

Kelly, J. C., Amerson, E. H., & Barth, J. T. (2012). Mild traumatic brain injury: Lessons learned from clinical, sports, and combat concussions. *Rehabilitation Research and Practice,* 5 pages.

Kirkendall, D. T., Jordan, S. E., & Garrett, W. E. (2001). Heading and head injuries in soccer. *Sports Medicine, 31,* 369–86.

Koepp, M. J., et al. (1998). Evidence for striatal dopamine release during a video game. *Nature, 393*(6682), 257–68.

Kolb, B., & Wishaw, I. (2003). *Fundamentals of Human Neuropsychology.* New York: W. H. Freeman and Co.

Kontos, A. P., Dolese, A., Elbin, R. J., Covassin, T., & Warren, B. L. (2011). Relationship of soccer heading to computerized neurocognitive performance and symptoms among female and male youth soccer players. *Brain Injury, 25,* 1234–41.

Kucuk, N. O., et al. (2000). Brain SPECT findings in long-term inhalant abuse. *Nuclear Medicine Communications, 21*(8), 769–73.

Kubota, K., et al. (1983). Effects of smoking on regional cerebral blood flow in neurologically normal subjects. *Stroke, 14*(5), 720–24.

Kumari, V., et al. (2004). Personality predicts brain responses to cognitive demands. *Journal of Neuroscience, 24*(47), 10636–41.

Kuo, L., Czarnecka, M., Kitlinska, J., Tilan, J., Kvetnansky, R., & Zukowska, Z. (2008). Chronic stress, combined with a high-fat/high-sugar diet, shifts sympathetic signaling toward neuropeptide Y and leads to obesity and the metabolic syndrome. *Annals of the New York Academy of Sciences, 1148,* 232–37.

Laatsch, L., et al. (1997). Impact of cognitive rehabilitation therapy on neuropsychological impairments as measured by brain perfusion SPECT: A longitudinal study. *Brain Injury, 11*(12), 851–63.

Lacerda, A., et al. (2003). Elevated thalamic and prefrontal regional cerebral blood flow in obsessive-compulsive disorder: A SPECT study. *Psychiatry Research, 123*(2), 125–34.

Lansing, K., Amen, D. G., & Hanks, C. (2005). High-resolution brain SPECT imaging and eye movement desensitization and reprocessing in police officers with PTSD. *Journal of Neuropsychiatry and Clinical Neurosciences, 17*(4), 526–32.

Levin, P., Lazrove, S., & van der Kolk, B. (1999). What psychological testing and neuroimaging tell us about the treatment of posttraumatic stress disorder by eye movement desensitization and reprocessing. *Journal of Anxiety Disorders, 13*(1–2), 159–72.

Liu-Ambrose, T., & Donaldson, M. (2009). Exercise and cognition in older adults: is there a role for resistance training programmes? *British Journal of Sports Medicine, 43*(1), 25–27.

Liu, J., et al. (2004). Malnutrition at age 3 years and externalizing behavior problems at ages 8, 11, and 17 years. *American Journal of Psychiatry, 161*(11), 150–54.

Lotfi, J., & Meyer, J. S. (1989). Cerebral hemodynamic and metabolic effects of chronic alcoholism. *Cerebrovascular and Brain Metabolism Review, 1*(1), 2–25.

Lovell, M. (2003). Recovery from mild concussion in high school athletes. *Journal of Neurosurgery, 98*(2), 296–301.

Lytle, M. E., et al. (2004). Exercise level and cognitive decline: The MoVIES Project. *Alzheimer's Disease and Associated Disorders, 18*(2), 57–64.

Martin, S. D., et al. (2001). Brain blood flow changes in depressed patients treated with interpersonal psychotherapy or venlafaxine hydrochloride: Preliminary findings. *Archives of General Psychiatry, 58*(7), 641–48.

Massagli, T., et al. (2004). Psychiatric illness after mild traumatic brain injury in children. *Archive of Physical Medicine and Rehabilitation, 85*(9), 1428–34.

Mathew, R. J., et al. (1997). Marijuana intoxication and brain activation in marijuana smokers. *Life Sciences, 60*(23), 2075–89.

Matthews, D. A. (1999). *Faith Factor.* New York: Penguin.

Mattson, M. P., Duan, W., & Guo, Z. (2003). Meal size and frequency affect neuronal plasticity and vulnerability to disease: Cellular and molecular mechanisms. *Journal of Neurochemistry, 84*(3), 417–31.

Mena, F. J., et al. (1998). Children Normal HMPAO Brain SPECT. *Alasbimn Journal, 1*(1).

Miller, A., & Coen, D. (1994). The case for music in the schools. *Phi Delta Kappan, 75*(6), 459–61.

Miller, B., & Cummings, J. (1998). *The Human Frontal Lobes.* New York: Guilford Press.

Miller, G. (2004). Society for Neuroscience meeting. Brain cells may pay the price for a bad night's sleep. *Science, 306*(5699), 1126.

Mueller, C., & Dweck, C. (1998). Praise for intelligence can undermine children's motivation and performance. *Journal of Personality and Social Psychology, 75*(1), 33–52.

Nakatani, E., et al. (2003). Effects of behavior therapy on regional cerebral blood flow in obsessive-compulsive disorder. *Psychiatry Research, 124*(2), 113–20.

Newberg, A. (2003). *Why God Won't Go Away.* New York: Ballantine Books.

Office of the Surgeon General Center for Mental Health Services. (1999). *Mental Health: A Report of the Surgeon General.* National Institute of Mental Health. http://www.surgeongeneral.gov/library/mentalhealth/home.html

Ophir, E., Nass, C. I., & Wagner, A. D. (2009). Cognitive control in media multitaskers. *Proceedings of the National Academy of Sciences of the United States of America, 106,* 15583–87.

Ortberg, J. (2003). *Everybody's Normal Till You Get to Know Them.* Grand Rapids, MI: Zondervan.

Paquette, V., et al. (2003). "Change the mind and you change the brain": Effects of cognitive behavior therapy on the neural correlates of spider phobia. *Neuroimaging, 18*(2), 401–09.

Riem, M. M., Bakermans-Kranenburg, M. J., van IJzendoorn, M. H., Out, D., & Rombouts, S. A. (2012). Attachment in the brain: Adult attachment representations predict amygdala and behavioral responses to infant crying. *Attachment and Human Development, 14*(6), 533–51. doi: 10.1080/14616734.2012.727252

Sala, M., et al. (2004). Stress and hippocampal abnormalities in psychiatric disorders. *European Neuropsychopharmacology, 14*(5), 393–405.

Sapolsky, R. M. (2001). Depression, antidepressants, and the shrinking hippocampus. *Proceedings of the National Academy of Science of the United States of America, 98*(22), 12320–22.

Sapolsky, R. M. (2004). *Why Zebras Don't Get Ulcers* (3rd ed.). New York: Owl Books.

Saxena, S., et al. (2001). Brain-behavior relationships in obsessive-compulsive disorder. *Seminars in Clinical Neuropsychiatry, 6*(2), 82–101.

Schlaug, G. (2001). The brain of musicians. A model for functional and structural adaptation. *Annals of the New York Academy of Sciences, 930,* 281–99.

Shi, X. Y. (2003). Cerebral perfusion SPECT imaging for assessment of the effect of hyperbaric oxygen therapy on patients with postbrain injury neural status. *Chinese Journal of Traumatology, 6*(6), 346–49.

Squeglia, L. M., Spadoni, A. D., Infante, M. A., Myers, M. G., & Tapert, S. F. (2009). Initiating moderate to heavy alcohol use predicts changes in neuropsychological functioning for adolescent girls and boys. *Psychology of Addictive Behaviors, 23*(4), 715–22. doi: 10.1037/a0016516

Stroth, S., Kubesch, S., Dieterle, K., Ruchsow, M., Heim, R., & Kiefer, M. (2009). Physical fitness, but not acute exercise modulates event-related potential indices for executive control in healthy adolescents. *Brain Research, 1269*(7), 114–24.

Sugiura, M., et al. (2000). Correlation between human personality and neural activity in cerebral cortex. *Neuroimage, 11*(5), 541–46.

Turner, R., et al. (2003). Brain function and personality in normal males: A SPECT study using statistical parametric mapping. *Neuroimage, 19*(3), 1145–62.

Volkow, N. D., et al. (1992). Decreased brain metabolism in neurologically intact healthy alcoholics. *American Journal of Psychiatry, 149*(8), 1016–22.

Waldman, M., Hochhauser, E., Fishbein, M., Aravot, D., Shainberg, A., & Sarne, Y. (2013). An ultra-low dose of tetrahydrocannabinol provides cardioprotection. *Biochemical Pharmacology, 85*(11), 1626–33. doi: 10.1016/j.bcp.2013.03.014. Epub 2013 Mar 26.

Wang, T., et al. (2008). Adverse effects of medical cannabinoids: A systematic review. *Canadian Medical Association Journal, 178*(13), 1669–78.

Williams, W. H., Cordan, G., Mewse, A. J., Tonks, J., & Burgess, C. N. (2010). Self-reported traumatic brain injury in male young offenders: A risk factor for re-offending, poor mental health and violence? *Neuropsychological Rehabilitation, 20*(6), 801–12. doi: 10.1080/09602011.2010.519613

Wu, J. C., Amen, D. G., & Bracha, S. (2000). Functional neuroimaging in clinical practice. In H. I. Kaplan & B. J. Sadock (Eds.), *The Comprehensive Textbook of Psychiatry*. Philadelphia: Lippincott Williams & Wilkins.

Yehuda, R., et al. (1993). Enhanced suppression of cortisol following dexamethasone administration in post-traumatic stress disorder. *American Journal of Psychiatry, 150*(1), 83–86.

http://pubs.niaaa.nih.gov/publications/aa63/aa63.htm

Zalesky, A., Solowij, N., Yucel, M., Lubman, D., Takagi, M., Harding, I., Lorenzetti, V., Wang, R., Searle, K., Pantelis, C., & Seal, M. (2012). Effect of long-term cannabis use on axonal fibre connectivity. *Brain, 135*(7), 2245–55.

Zellner, D., Loaiza, S., Gonzalez, Z., Pita, J., Morales, J., Pecora, D., & Wolf, A. (2006). Food selection changes under stress. *Physiology & Behavior, 87*(4), 789–93.

Index

dementia, 40, 44
depression, 14, 16, 40, 44, 215
    appetite and, 61
    brain scan, *129*
    breaking the cycle, 67
    example, deep limbic system and, 63–65
    exercise and alleviating, 207–8
    men vs. women with, 67
    number of children, adolescents and adults with, 17, 207
    overactive deep limbic system and, *60,* 60, 61, 63, 65, 78
    post-concussion syndrome and, 104
    seeking help for, 131
    self-medicating, 96–99
    stress and, 228, 229
diaphragmatic breathing, 233–34
diet soda, 198
distractibility, 38, 39, 44, 78, 97, 112, 120, 123, 125, 142, 214. *See also* ADD/ADHD; attention span and focus
dopamine, 95–96, 222
drug use, *14, 86,* 88–93, *92*
    before and after age twenty-five, 87–88
    brain damage and, 85–88
    individual reactions to, 99–100
    questions to answer about, 101
    reasons for using, 94–99
    as self-medication, 96–99, 100
    summary, 100–101
dysthymia, 65

E

eating disorders, 17, 54, 57
Ecstasy, 87, 100
emotions
    amygdala and, 28
    controlling, 37
    deep limbic system and, 59–69, 96
    memories and, 62
    teenage decision-making and, 29–30
    teenage years, 28, 59
    "Thinking with Your Feelings" ANT, 179–80
empathy, 36, 37, 38, 39
environmental toxins, *107,* 107–8
    inhalants, 108
exercise, mental, 213–25. *See also* Two-Week Brain Smart Plan
    challenging yourself, 220
    coordination games and activities, 222
    creative games (right brain), 221
    games to stimulate your brain, 220–23
    memory games, 223
    music games, 222
    perfect practice and, 218–19
    questions to answer about, 225
    strategy games, 222–23
    summary, 225
    10 ways to exercise your brain, 223–24
    visual games (left brain), 223
    word games, 221
exercise, physical, 205–11. *See also* Two-Week Brain Smart Plan
    academic performance and, 206–7
    addiction and lowering cravings, 208
    anxiety and stress helped by, 208
    cardiovascular (aerobic), 209
    combo exercises, 210
    coordination exercises, 209–10
    depression helped by, 207–8
    Exercise Plan worksheet, 246–48
    memory boosting with, 207

organization skills, 12, 17, 34, 38, 39, 78

oxygen deprivation, 109

"choking game" and, 109

P

pain, chronic, 54, 57, 65

panic disorder, 76, 208

parietal lobes, 15, *26*, 27, 222

Parkinson's disease, 72

part-time jobs, 218

peer pressure, 29–31, 96

Temple University study, 30–31

perfectionism, 77

personality disorders, 17

phobias, 54

planning skills, 12, 36, 38

PMS (premenstrual syndrome), 57

prefrontal cortex, *15*, 28, 29, 30, *33*, 33–45, 77–78, 125, 126, 131–32, 139, 214, 237, 239

author's father and, 120

conditions associated with, 39–40, 44

dealing with others who have prefrontal cortex issues, 150–52

development of, 31

example, elementary school student, 40–41

exercises to alleviate problems, 210

forethought, planning, judgment, and, 32, 34, 35–36, 37, 38, 40, 42, 43, 44, 52, 96, 132, 141, 150

functions, 33–34, 35–37

injury to, 12

low activity, positive traits and, 44

meditation/prayer, positive effects of, 232–33

One-Page Miracle as booster, 243

problems with, 37–39, *38*

questions to answer about, 45

responding to lack of development, 42–43

size of, compared to animal brains, 34

supervision vs. lack of supervision and, 43–44

treatment for low activity of, 45

procrastination, 38, 73, 78, 214

PTSD (post-traumatic stress disorder), 57, *72*, 72, 76, 106, 229

R

relationships. *See also* Two-Week Brain Smart Plan

with bosses, 151–52, 157–59

dealing with others who have cingulate issues, 149–50, 152–59

dealing with others who have deep limbic system and basal ganglia issues, 159–61

dealing with others who have prefrontal cortex issues, 150–52

deep limbic system and, 62, 63, 215

with friends and significant others, 160–61

knowing other people's brains and, 137–48, 259

with parents, 149–50, 152–53, 160

questions to answer about, 148, 162

sample conversation, teenager and parents with overactive cingulate, 153

Step 1: Understand the Basics of Brain Function, 141–43

Step 2: Know Your Own Brain, 143–45, 239

Step 3: Accept That You Cannot Change Others, 145–47

Visit
www.brain25.com
for tons of free information,
downloads and resources to help you
change your brain and change your life.